Geometry Theorems

Xing Zhou

Math for Gifted Students

http://www.mathallstar.org

Copyright © 2015 by Xing Zhou. All rights reserved.

No part of this book may be reproduced, distributed or transmitted in any form or by any means, including photocopying, scanning, or other electronic or mechanical methods, without written permission of the author.

To promote education and knowledge sharing, reuse of some contents of this book may be permitted, courtesy of the author, provided that: (1) the use is reasonable; (2) the source is properly quoted; (3) the user bears all responsibility, damage and consequence of such use. The author hereby explicitly disclaims any responsibility and liability; (4) the author is notified in advance; and (5) the author encourages, but does not enforce, the user to adopt similar policies towards any derived work based on such use.

Please visit the website `http://www.mathallstar.org` for more information or email `contact@mathallstar.org` for suggestions, comments, questions and all copyright related issues.

use your mobile device to scan this QR code for more resources including books, practice problems, online courses, and blog.

This book was produced using the LaTeX system.

Contents

1. **Introduction** 1
   1.1 Learning Guide 1
   1.2 Conventions 2
2. **Right Triangle** 3
   2.1 Pythagorean Theorem 3
   2.2 Pythagorean Triplet 5
   2.3 Useful Conclusions 7
   2.4 Practice 9
3. **Arbitrary Triangle** 13
   3.1 Angle Bisector Theorem 13
   3.2 Law of Sines and Law of Cosines 15
   3.3 Stewart Theorem 18
   3.4 Practice 20
4. **Circle** 23
   4.1 Central Angle and Inscribed Angle 23
   4.2 Concyclic Quadrilateral 25
   4.3 Ptolemy's Theorem 29
   4.4 Power of a Point 31
   4.5 Circumscribed Quadrilateral 33
   4.6 Practice 35
5. **Cevian and Transversal** 39
   5.1 Cevian and Transversal Defined 39
   5.2 Ceva's Theorem 40
   5.3 Menelaus' Theorem 41
   5.4 Trigonometric Ceva's Theorem 42
   5.5 Concurrent Lines 43
   5.6 Practice 44
6. **Additional Topics** 49
   6.1 Area Computation 49

CONTENTS

	6.1.1	Triangle	49
	6.1.2	Ratio of Triangles Areas	51
	6.1.3	Quadrilateral	55
	6.1.4	Pick's Theorem	56
6.2	Triangle Centers	57
	6.2.1	Centroid (Center of Mass)	57
	6.2.2	Orthocenter	58
	6.2.3	Circumcenter	61
	6.2.4	Incenter	62
6.3	Special Points, Lines and Others	64
6.4	Practice	65

7 Solutions 71
 7.1 *Chapter 1* 72
 7.2 *Chapter 2* 73
 7.3 *Chapter 3* 83
 7.4 *Chapter 4* 91
 7.5 *Chapter 5* 105
 7.6 *Chapter 6* 118

Preface

Welcome to Math All Star© series!

Math All Star originates from a series of lectures given to a group of gifted middle school students with a love for mathematics and an interest in participating in competitions such as MathCounts, AMC, and AIME. These lectures aim to strengthen their problem-solving abilities and to introduce effective techniques that are not typically taught in the classroom.

As the popularity of Math All Star grew, the author began to upload lecture materials to create online courses, thereby providing students with the opportunity to progress at their own paces.

Since then, course materials have constantly been reviewed and updated to reflect student feedback and the observations made during lectures. Recent competition problems are also continuously analyzed and referenced to ensure the relevance of the contents. These course materials are the foundations of this Math All Star series.

Because competition math is a diversified subject that covers both a wide breadth and depth of topics, it is quite challenging to effectively cover all the material in one book that is appropriate for every interested student. Consequently, the author has decided to write a series of books, with each one focusing on a particular topic. Students are encouraged to pick and choose where to begin, depending on their individual skill levels and needs.

CONTENTS

In addition to these books, the Math All Star website provides extra practice problems and serves as a highly recommended supplemental learning resource.

If there are any questions, comments, or concerns, please visit the website or email **contact@mathallstar.org**.

Happy learning!

To visit the Math All Star website, scan this QR code or go directly to
http://www.mathallstar.org

Chapter 1

Introduction

1.1 Learning Guide

Solving competition geometry problems is challenging. It requires students to be familiar with many geometry theorems as well as to be proficient in employing various techniques. These two skills are complementary to each other. This book focuses on the theorem part. *Geometry Techniques*, a book written by the same author, covers the technique part.

There are hundreds of, if not more, geometry theorems. It is nearly impossible and definitely not necessary to remember all of them. Instead, students should focus on those must know theorems. In addition to remembering these theorems themselves, it is also important to study their typical applications. This book covers both topics.

Each chapter of this book introduces a collection of related theorems. Those essential ones are discussed in the body contents. Students must remember all of them. Some additional theorems are included in the practices. They are good to know and remember.

Chapter 1: Introduction

Practices are intended to demonstrate these theorems' typical applications. Many of them are classical problems. Some conclusions are well-known and can be practically treated as theorems. As such, students are encouraged to remember such conclusions as well.

An index of discussed theorems and formulas is provided at the end of this book.

1.2 Conventions

In order to keep description concise, following conventions will be followed in this book unless specifically mentioned otherwise.

In $\triangle ABC$:

(i) A, B, and C: both angles and vertices

(ii) a, b, and c: opposite sides to A, B, and C, respectively

(iii) S: this triangle's area

(iv) p: semi-perimeter, i.e. $p = \frac{1}{2}(a + b + c)$

(v) R: circumradius, the radius of $\triangle ABC$'s circumcircle

(vi) r: inradius, the radius of $\triangle ABC$'s inscribed circle

(vii) h_a, h_b, and h_c: altitudes drawn from A, B, and C

(viii) m_a, m_b, and m_s: medians drawn from A, B, and C

Chapter 2

Right Triangle

Let's start our journey by looking into right triangle related theorems.

2.1 Pythagorean Theorem

Undoubtedly, Pythagorean theorem is one of the best known geometry theorems.

> **Theorem 2.1.1 Pythagorean Theorem**
>
> Let $\triangle ABC$ be a right triangle and $\angle C = 90°$. Pythagorean theorem states that the following relation always holds:
> $$AB^2 = AC^2 + BC^2$$

The converse of the Pythagorean theorem also holds: if $AB^2 = AC^2 + BC^2$, then $\triangle ABC$ is a right triangle where $\angle C = 90°$. This conclusion is often used to determine whether $\triangle ABC$ is a right triangle.

Chapter 2: Right Triangle

> When studying geometry theorems, it is important and useful to investigate their converses. Some converses do not necessarily hold, but many do.

Proofs of Pythagorean theorem and its converse can be found in many books and materials. Therefore, they will not be discussed here.

Pythagorean theorem has several generalizations and extensions. For example:

(i) What if the angles is not 90°?

(ii) Is there a similar conclusion in 3-dimensional space?

The answer to the 1st question is Law of Cosines which will be discussed in next chapter. The answer to the 2nd question is De Gua's theorem.

> **Theorem 2.1.2 De Gua's Theorem**
>
> If three faces of a tetrahedron are right triangles whose right angles share a same vertex[a], then the square of this tetrahedron's base's area equals the sum of squares of its three sides' areas.
>
> [a]Such a tetrahedron is sometime referred as a trirectangular tetrahedron.

$$S^2_{\triangle BCD} = S^2_{\triangle ABC} + S^2_{\triangle ACD} + S^2_{\triangle ADB}$$

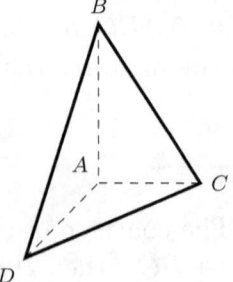

De Gua's theorem is reasonably intuitive and can be proved using Pythagorean theorem. Its proof will be left as a practice.

Chapter 2: Right Triangle

2.2 Pythagorean Triplet

Among all right triangles, those with integer side lengths often attract additional interests. The smallest such triangle is $3-4-5$. Others include $6-8-10$, $5-12-13$, and so on. In some cases, problems related to such triangles can be solved by using Pythagorean theorem directly. But quite often, applying Pythagorean triplet formula can be a more convenient way.

Pythagorean triplet formula gives a general solution to the following equation where a, b, and c are all positive integers.

$$a^2 + b^2 = c^2 \tag{2.1}$$

An integer triplet (a, b, c) satisfying *(2.1)* is called a Pythagorean triplet because they can be lengths of a right triangle's three edges.

> **Theorem 2.2.1 Pythagorean Triplet Formula**
>
> All Pythagorean triplets can be generated using the following formula by choosing appropriate positive integers m and n satisfying $m > n$.
>
> $$\begin{cases} a = m^2 - n^2 \\ b = 2mn \\ c = m^2 + n^2 \end{cases} \tag{2.2}$$

For example, letting $m = 2$ and $n = 1$ leads to the smallest Pythagorean triplet $(3, 4, 5)$. Similarly, $(5, 12, 13)$ can be obtained by setting $m = 3$ and $n = 2$.

Pythagorean theorem and Pythagorean triplet formula are two primary tools to solve right triangle related problems. Let's consider the following example.

Chapter 2: Right Triangle

Example 2.2.1

Find all right triangles with integer side lengths whose perimeter and area are numerically equal.

By applying the Pythagorean theorem directly, the following equation can be obtained:
$$\frac{1}{2}ab = a + b + \sqrt{a^2 + b^2}$$

This is an indeterminate equation[1] which is solvable. However in this particular case, applying the Pythagorean triplet formula can lead to a simpler equation to solve.

Solution

By *(2.2)*, we have
$$\frac{1}{2}ab = a + b + c$$
$$\frac{1}{2} \cdot (m^2 - n^2) \cdot 2mn = (m^2 - n^2) + 2mn + (m^2 + n^2)$$
$$mn(m+n)(m-n) = 2m(m+n)$$
$$n(m-n) = 2$$

Because m and n are both positive integers, this above equation can hold if and only if
$$\begin{cases} n = 1 \\ m - n = 2 \end{cases} \text{ or } \begin{cases} n = 2 \\ m - n = 1 \end{cases}$$

This leads to two solutions: $(m, n) = (3, 1)$ or $(3, 2)$. This implies there are only two such right triangles: $(6, 8, 10)$ and $(5, 12, 13)$.

Done.

---

[1] Solving various types of indeterminate equations is an important skill to master. This is covered in the book *Indeterminate Equation* by the same author.

2.3 Useful Conclusions

Right triangles have many interesting properties. Some of them are listed here.

<u>Geometric Mean Theorem</u>

Let CD be an altitude of $\triangle ABC$ where $\angle C = 90°$. Then the following relations always hold:

$$CD^2 = AD \cdot BD$$
$$AC^2 = AD \cdot AB$$
$$BC^2 = BD \cdot AB$$

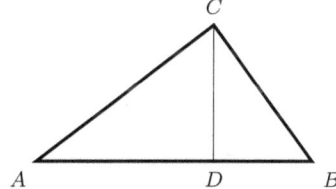

The geometric mean theorem can be proved by noticing that

$$\triangle ACD \sim \triangle ABC \sim \triangle CBD$$

<u>Circumradius</u>

$$R = \frac{c}{2}$$

In particular, a right triangle's circumcenter is the midpoint of its hypotenuse.

<u>Inradius</u>

$$r = \frac{a+b-c}{2} \tag{2.3}$$

Inradius is often related to area computation. But in the case of a right triangle, it can be computed directly. This is shown in the following example. A more general formula to compute inradius r of an arbitrary triangle will be given in next chapter.

Chapter 2: Right Triangle

Example 2.3.1

Given a right triangle, show that
$$r = \frac{a+b-c}{2}$$

Proof

As shown,

$$\begin{cases} r + x = a \\ r + y = b \\ x + y = c \end{cases}$$

$$\therefore r = \frac{a+b-c}{2}$$

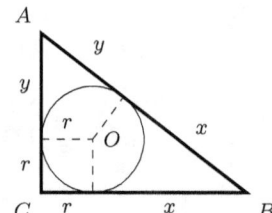

QED

The next example is an important conclusion which is often found in medium to advanced level competitions.

Example 2.3.2

Given two segments AB and MN, show that
$$MN \perp AB \Leftrightarrow AM^2 - BM^2 = AN^2 - BN^2$$

Proof

Let's first solve the easier part: if $MN \perp AB$, then $AM^2 - BM^2 = AN^2 - BN^2$.

Suppose MN and AB intersect at point O.

∵ $AM^2 - AN^2$
$= (AM^2 - AO^2) - (AN^2 - AO^2)$
$= MO^2 - NO^2$

$BM^2 - BN^2$
$= (BM^2 - BO^2) - (BN^2 - BO^2)$
$= MO^2 - NO^2$

∴ $AM^2 - AN^2 = BM^2 - BN^2 \implies AM^2 - BM^2 = AN^2 - BN^2$

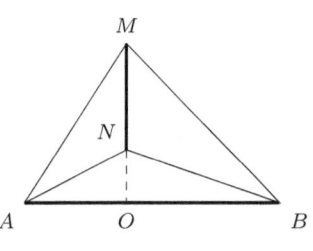

The 2nd part of the proof will be discussed in the next chapter after Law of Cosines has been introduced.

QED

2.4 Practice

Practice 1

(**Hippocrates Problem**) As shown, three semi-circles are drawn on three sides of right $\triangle ABC$. Show that the sum of shaded areas equals the area of $\triangle ABC$.

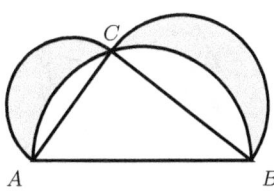

Practice 2

Given a rectangular prism with side lengths of 3, 4, and 5, as shown, what is the length of the shortest route from A to C' via surfaces.

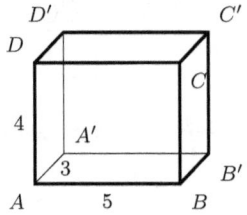

Practice 3

(British Flag Theorem) Let point P lie inside rectangle $ABCD$. Draw four squares using each of AP, BP, CP, and DP as one side. Show that

$$S_{AA_1A_2P} + S_{CC_1C_2P} = S_{BB_1B_2P} + S_{DD_1D_2P}$$

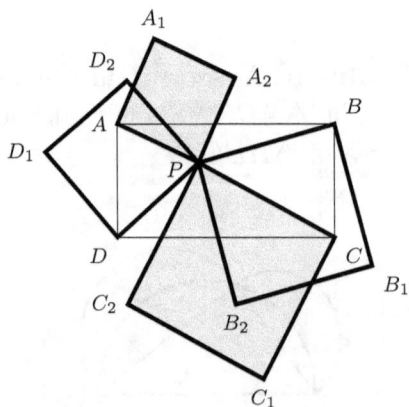

Chapter 2: Right Triangle

Practice 4

(**De Gua's Theorem**) In a trirectangular tetrahedron $ABCD$ where A is the shared right-angle corner. Show that

$$S^2_{\triangle BCD} = S^2_{\triangle ABC} + S^2_{\triangle ACD} + S^2_{\triangle ADB}$$

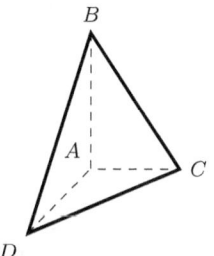

Practice 5

Let CD be the altitude in right $\triangle ABC$ from the right angle C. If inradii of $\triangle ABC$, $\triangle ACD$, and $\triangle BCD$ are r_1, r_2, and r_3, respectively, show that

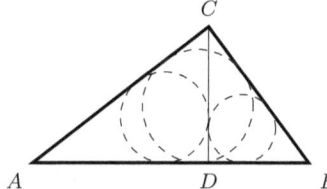

$$r_1 + r_2 + r_3 = CD$$

Practice 6

Let $\triangle ABC$ be a right triangle where $\angle C = 90°$. Show that if point D is on side BC,

$$AB^2 = DB^2 + DA^2 + 2 \cdot DB \cdot DC$$

If D locates on BC's extension, then

$$AB^2 = DB^2 + DA^2 - 2 \cdot DB \cdot DC$$

Chapter 2: Right Triangle

Practice 7

Three circles are tangent to each other and also a common line, as shown. Let the radii of circles O_1, O_2, and O_3 be r_1, r_2, and r_3, respectively. Show that

$$\frac{1}{\sqrt{r_3}} = \frac{1}{\sqrt{r_1}} + \frac{1}{\sqrt{r_2}}$$

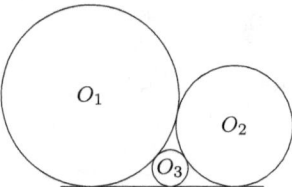

Practice 8

Let point P be inside an equilateral $\triangle ABC$ such that $AP = 3$, $BP = 4$, and $CP = 5$. Find the side length of $\triangle ABC$.

Practice 9

Let M be a point inside $\triangle ABC$. Draw $MA' \perp BC$, $MB' \perp CA$, and $MC' \perp AB$ such that $BA' = BC'$ and $CA' = CB'$. Prove $AB' = AC'$.

(Ref 1979 China)

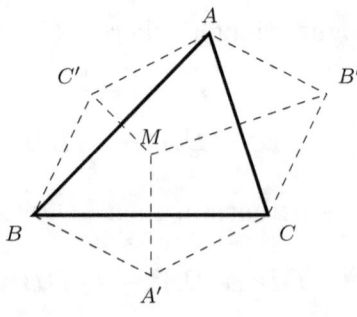

Chapter 3

Arbitrary Triangle

This chapter covers several important theorems that hold for any triangle. Please note that both Ceva's theorem and Menelaus' theorem hold for any triangle too. But they will be discussed in *Chapter 5 Cevian and Transversal*.

3.1 Angle Bisector Theorem

This theorem holds for both interior angle bisector and exterior angle bisector.

> **Theorem 3.1.1 Angle Bisector Theorem**
>
> In $\triangle ABC$, let AD bisect $\angle A$ (or its exterior angle) and meet BC (or its extension) at D. Angle bisector theorem states that the following relation always holds
>
> $$\frac{AB}{AC} = \frac{BD}{DC} \qquad (3.1)$$

Chapter 3: Arbitrary Triangle

Please note that, regardless of AD is an interior angle bisector or an exterior one, the expression of *(3.1)* stays the same. Its right side is always the ratio between the two segments each of which is formed by one vertex and the section point.

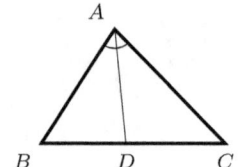
Interior Angle Bisector Theorem

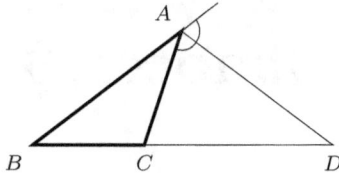
Exterior Angle Bisector Theorem

Interior angle bisector theorem can be proved by using similar triangles. Draw $A'C \parallel AB$ meeting AD's extension at A'.

$$\angle CA'D = \angle BAD$$
$$\angle ADB = \angle A'DC$$
$$\Longrightarrow \triangle ABD \sim \triangle A'CD$$
$$\Longrightarrow \frac{AB}{A'C} = \frac{BD}{DC}$$

$$\angle CA'D = \angle BAD = \angle CAD \Longrightarrow AC = A'C \Longrightarrow \frac{AB}{AC} = \frac{BD}{DC}$$

Exterior angle bisector theorem can be proved by using a similar technique: draw $A'C \parallel AD$ meeting AB at A'.

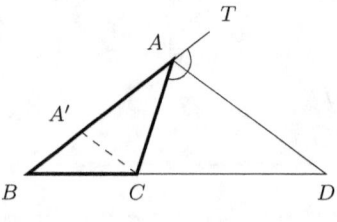

$$A'C \parallel AD \Longrightarrow \frac{BD}{CD} = \frac{AB}{AA'}$$
$$\angle AA'C = \angle TAD = \angle DAC = \angle ACA'$$
$$\Longrightarrow AA' = AC$$
$$\therefore \frac{BD}{CD} = \frac{AB}{AC}$$

Chapter 3: Arbitrary Triangle

3.2 Law of Sines and Law of Cosines

These two theorems establish general relations among a triangle's sides and angles. In addition, they often serve as stepping stones for employing trigonometric identities and transformation to solve geometry problems.

> **Theorem 3.2.1 Law of Sines**
>
> Given any $\triangle ABC$, the following relation will hold.
> $$\frac{a}{\sin A} = \frac{b}{\sin B} = \frac{c}{\sin C} = 2R \qquad (3.2)$$
> where R is $\triangle ABC$'s circumradius.

The validity of this theorem can be proved by observing $\triangle ODC$ in the diagram below. By central angle theorem[1], we have $\angle BOC = 2\angle A$ which implies $\angle DOC = \angle A$. It follows that

$$CD = OC \cdot \sin \angle A \implies \frac{a}{2} = R \sin \angle A \implies \frac{a}{\sin A} = 2R$$

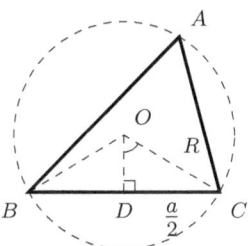

One application of Law of Sines is to convert between ratios of sides and ratios of angles' sine values. For example, if $\sin \angle A : \sin \angle B : \sin \angle C = 3 : 4 : 5$, then we can assert that $\triangle ABC$ is a right triangle because $a : b : c = 3 : 4 : 5$.

---

[1]The central angle theorem will be discussed in *Chapter 4*.

Chapter 3: Arbitrary Triangle

While Law of Sines establishes relationship among side, angle and circumradius, Law of Cosines can be thought of an extension to the Pythagorean theorem when the angle is not a right angle.

> **Theorem 3.2.2 Law of Cosines**
>
> Given any $\triangle ABC$, the following relations always hold.
>
> $$a^2 = b^2 + c^2 - 2bc\cos A$$
> $$b^2 = c^2 + a^2 - 2ca\cos B$$
> $$c^2 = a^2 + b^2 - 2ab\cos C$$

Clearly, when the corresponding angle is a right angle, Law of Cosines will become Pythagorean theorem due to the fact that $\cos 90° = 0$.

There are several ways to prove this theorem. One of them is based on Pythagorean theorem.

Proof

Without loss of generality, let's show $a^2 = b^2 + c^2 - 2bc\cos A$. In order to derive this result, let's draw an altitude from B meeting AC at D.

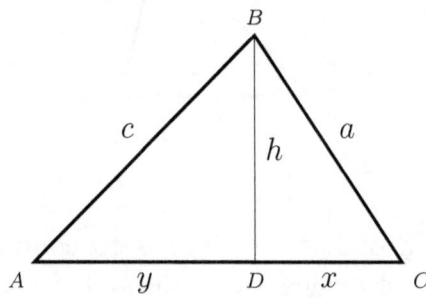

Notice that $x + y = b$, we have

$$\begin{aligned} a^2 &= x^2 + h^2 \\ &= (b-y)^2 + h^2 \\ &= b^2 - 2by + y^2 + h^2 \\ &= b^2 - 2bc\cos A + c^2 \end{aligned}$$

QED

Having studied Law of Cosines, we are now ready to complete the proof of *Example 2.3.2* on *page 8*

Given two segments AB and MN, show that

$$AM^2 - BM^2 = AN^2 - BN^2 \implies MN \perp AB$$

Proof

Suppose MN and AB intersect at point O. Let $\angle MOA = \alpha$, $\angle MOB = \beta$. Then the claim is equivalent to proving

$$AM^2 - AN^2 = BM^2 - BN^2$$
$$\implies \alpha = \beta = 90°$$

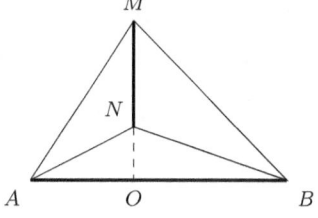

By Law of Cosines:

$$\begin{aligned} AM^2 - AN^2 &= (AO^2 + MO^2 - 2AO \cdot MO\cos\alpha) \\ &\quad - (AO^2 + NO^2 - 2AO \cdot NO\cos\alpha) \\ &= MO^2 - NO^2 - 2AO(MO - NO)\cos\alpha \end{aligned}$$

Chapter 3: Arbitrary Triangle

and

$$BM^2 - BN^2 = (BO^2 + MO^2 - 2BO \cdot MO \cos \beta)$$
$$- (BO^2 + NO^2 - 2BO \cdot NO \cos \beta)$$
$$= MO^2 - NO^2 - 2BO(MO - NO) \cos \beta$$

$$\therefore \quad AO \cos \alpha = BO \cos \beta$$

Because both α and β are positive, and $\alpha + \beta = 180°$, the above equality can only hold if both sides equal 0, or $\alpha = \beta = 90°$. Otherwise, $\cos \alpha$ and $\cos \beta$ will have opposite signs which means two sides cannot be equal.

QED

3.3 Stewart Theorem

Stewart theorem describes relation among a triangle's cevian[2] and three sides. Some of its direct applications are to compute the length of medians, angle bisectors, altitudes, and so on.

Theorem 3.3.1   Stewart Theorem

Given a triangle as shown on the right where each letter represents the length of its corresponding segment, then

$$b^2 m + c^2 n = a(d^2 + mn)$$

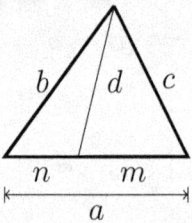

---

[2]The concept of cevian is explained in *Section 5.1* on *page 39*.

Proof

Denote the angle between the line segments d and m as θ. Then by the Law of Cosines, we have:
$$c^2 = d^2 + m^2 - 2dm\cos\theta$$
$$b^2 = d^2 + n^2 - 2dn\cos(180° - \theta)$$

Notice that $\cos(180° - \theta) = -\cos\theta$, we have
$$\begin{aligned}b^2m + c^2n &= m(d^2 + n^2 + 2dn\cos\theta) + n(d^2 + m^2 - 2dm\cos\theta)\\&= md^2 + mn^2 + nd^2 + nm^2\\&= (m+n)d^2 + (m+n)mn\\&= (m+n)(d^2 + mn)\\&= a(d^2 + mn)\end{aligned}$$

QED

Let's use the Stewart theorem to compute the length of an angle bisector. Computing the length of a median is left as a practice. It can be solved in a similar way.

Example 3.3.1

Let AD bisect $\angle A$ in $\triangle ABC$, show that

$$d^2 = \frac{bc}{(b+c)^2}\left((b+c)^2 - a^2\right)$$

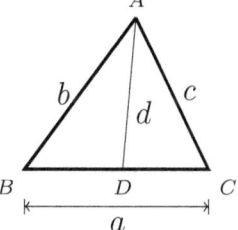

The strategy to solve such type of problems is first to use the properties of the involved cevian (e.g. angle bisector in this case) to establish certain relation. The next step is to set such relationship to Stewart theorem formula.

Proof

Applying angle bisector theorem:

$$m = \frac{ab}{b+c} \quad \text{and} \quad n = \frac{ac}{b+c}$$

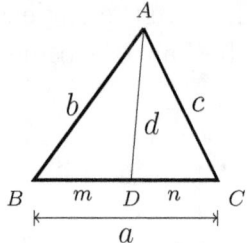

Then, apply Stewart theorem:

$$b^2 \frac{ac}{b+c} + c^2 \frac{ab}{b+c} = a^2 \left(\frac{ab}{b+c} \cdot \frac{ac}{b+c} + d^2 \right)$$

Rearranging the above equation yields the to-be-proved claim.

QED

3.4 Practice

Practice 1

Given a triangle, show that the angle formed by the angle bisector and the altitude from the same vertex equals half of the difference between the two base angles.

Practice 2

(**Apollonius' Theorem**) Let AD be one median of $\triangle ABC$ where point D lies on side BC. Show that the following relation holds:
$$AB^2 + AC^2 = 2 \times (AD^2 + BD^2)$$

Chapter 3: Arbitrary Triangle

Practice 3

Show that when $\triangle ABC$ is a right triangle, Apollonius' theorem (see the previous practice problem) will reduce to Pythagorean theorem.

Practice 4

In a right triangle whose legs are a and b, and hypotenuse is c, two segments drawn from the right angle divide the hypotenuse into three equal parts of length x. If the lengths of these two segments are p and q, show that $p^2 + q^2 = 5x^2$.

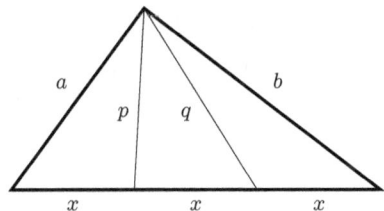

Practice 5

Given a parallelogram $ABCD$, show that

$$AB^2 + BC^2 + CD^2 + DA^2 = AC^2 + BD^2$$

Practice 6

(**Law of Tangents**) In any triangle, show that

$$\frac{a+b}{a-b} = \frac{\tan(\frac{A+B}{2})}{\tan(\frac{A-B}{2})}$$

Chapter 3: Arbitrary Triangle

Practice 7

Find the measure of the largest interior angle of $\triangle ABC$ if its side lengths are 3, 5, and 7.

Practice 8

If circle O has an inscribed pentagon whose sides lengths are 3, 3, 5, 5 and 7, in that order. Find the area of this circle.

Practice 9

Given $\triangle ABC$, $\angle C$ is quadrisected (divided into four equal angles) by the altitude, the angle bisector, and the median from that vertex C. Find the measurement of $\angle C$.

Practice 10

(Law of Cosines in 3-Dimensional Space) In tetrahedron $ABCD$, let the areas of $\triangle BCD$, $\triangle ACD$, $\triangle ABD$, and $\triangle BCD$ be a, b, c, and d, respectively. Also let (a,b), (b,c), and (c,a) be the angles between faces $DBC - DCA$, $DCA - DAB$, and $DAB - DBC$, respectively. Prove

$$d^2 = a^2 + b^2 + c^2 - 2ab\cos(a,b) - 2bc\cos(b,c) - 2ca\cos(c,a)$$

Chapter 4

Circle

Circle is an important and well-studied geometric shape. It has a rich set of properties. We will focus on those that are most relevant to middle school and high school math competitions.

4.1 Central Angle and Inscribed Angle

Given a circle, a central angle is an angle whose vertex is the circle's center and whose legs are two radii. Correspondingly, an inscribed angle is an angle which is formed by two chords sharing a common endpoint.

> **Theorem 4.1.1 Central Angle Theorem**
>
> An inscribed angle is always equal to half of a corresponding central angle. By extension, if two inscribed angles are subtended from the same arc, they must be congruent.
>
>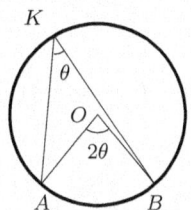

Chapter 4: Circle

In order to show that $\angle AOB = 2\angle K$, we can connect KO and extend it to meet circle O at C. Because AO, BO, and KO are all radii, they must equal in lengths. This means both $\triangle AOK$ and $\triangle BOK$ are isosceles.

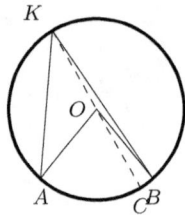

$$\angle AOC = \angle OAK + \angle OKA = 2\angle OKA$$
$$\angle BOC = \angle OBK + \angle OKB = 2\angle OKB$$

$\therefore \quad \angle AOB = \angle AOC + \angle BOC = 2(\angle OKA + \angle OKB) = 2\angle K$

Let's review another example.

Example 4.1.1

Let AX bisect $\angle A$ and intersect $\triangle ABC$'s circumcircle at X. Point I locates on AX. Prove that I is the incenter of $\triangle ABC$ if and only if $BX = CX = BI$.

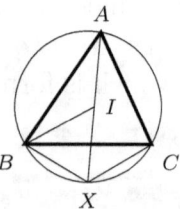

Proof

First, $BX = CX$ is obvious because AX bisects $\angle A$. Then

I is incenter

$\Leftrightarrow \angle CBI = \angle ABI$

$\Leftrightarrow \angle IBX - \angle CBX = \angle BIX - \angle BAX$

$\Leftrightarrow \angle IBX = \angle BIX$

$\Leftrightarrow BI = BX$

QED

4.2 Concyclic Quadrilateral

Concyclic quadrilaterals appear often in geometry problems. There are two basic criteria to determine whether a quadrilateral is concylic.

Theorem 4.2.1   Concyclic Quadrilateral

A quadrilateral $ABCD$ is concyclic if any of the following criteria is met:

i) $\angle ABC + \angle CDA = 180°$ or $\angle BCD + \angle DAB = 180°$

ii) $\angle DAC = \angle DBC$ (or equivalent expressions)

Conversely, if $ABCD$ is concyclic, then these angle relationships will hold.

Example 4.2.1

Four sides of a concyclic quadrilateral have lengths of 25, 39, 52, and 60, in that order. Find the circumference of its circumcircle.
(Ref 1995 China)

Chapter 4: Circle

Solution

Because $ABCD$ are concyclic, $\angle A + \angle C = 180°$ which implies

$$\cos \angle A = -\cos \angle C$$

Applying Law of Cosines on $\triangle ABD$ and $\triangle CBD$, respectively:

$$\begin{cases} BD^2 = AB^2 + AD^2 - 2AB \cdot AD \cos \angle A \\ BD^2 = CB^2 + CD^2 - 2CB \cdot CD \cos \angle C \end{cases}$$

Setting $\cos \angle C = -\cos \angle A$ and all the given conditions into the above system leads to

$$\cos \angle A = \frac{25^2 + 60^2 - 39^2 - 52^2}{2 \times (25 \times 60 + 29 \times 52)} = 0 \implies \angle A = 90°$$

This means BD is the diameter. It follows

$$BD = \sqrt{25^2 + 60^2} = 65$$

Hence the answer is 65π.

Done.

Some problems are related to constructing three outwards triangles on an arbitrary triangle's sides. *Napoleon's triangle*[1] is one of the most famous examples. These problems are often related to concyclic quadrilaterals.

First, let's investigate the condition when the three new triangles' circumcircles will meet at a single point.

---

[1]Napoleon's triangle will be discussed shortly.

Chapter 4: Circle

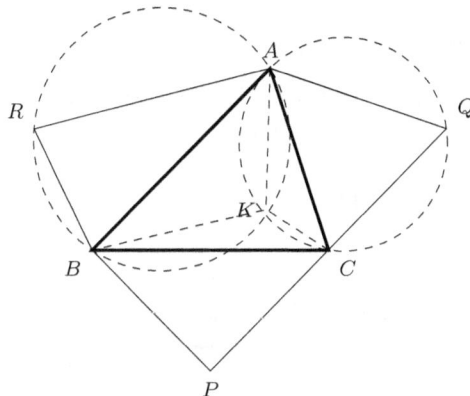

As shown above, let circumcircles ABR and ACQ meet K in addition to A. The goal is to determine the condition when $KBPC$ is concyclic.

$$\begin{aligned}\angle BKC &= 360° - \angle AKB - \angle AKC \\ &= 360° - (180° - \angle R) - (180° - \angle Q) \\ &= \angle R + \angle Q\end{aligned}$$

For $KBPC$ to be concyclic, sum of $\angle BKC + \angle P$ must be 180°. This implies that

$$\angle P + \angle Q + \angle R = 180° \qquad (4.1)$$

There exist many different scenarios under which relation *(4.1)* is satisfied. One case is that the three outbound triangles are similar where $\angle P$, $\angle Q$, and $\angle R$ represent three different angles. In such a case, it can be shown that the triangle formed by their circumcenters is similar to them too. This is illustrated in the next example.

Chapter 4: Circle

Example 4.2.2

As shown, let O_1, O_2, and O_3 be circumcenters of $\triangle PBC$, $\triangle QCA$, and $\triangle RAB$, respectively. If

$$\triangle ABR \sim \triangle CPB \sim \triangle QCA$$

Prove

$$\triangle O_2 O_1 O_3 \sim \triangle ABR \sim \triangle CPB \sim \triangle QCA$$

Solution

Because $\triangle ABR \sim \triangle CPB \sim \triangle QCA$, it hold that

$$\angle P + \angle Q + \angle R = 180°$$

Therefore the three circumcircles are concurrent. Suppose they meet at point K. Because $O_2 O_3$ connects two circle centers and AK is a shared chord, it must hold that $O_2 O_3 \perp AK$. Similarly, $O_1 O_2 \perp CK$ and $O_1 O_3 \perp BK$.

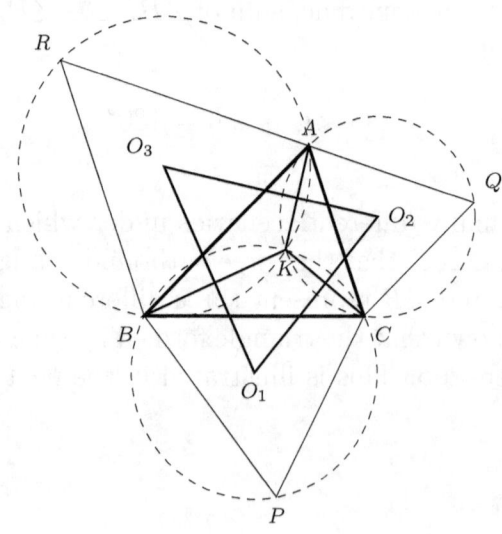

Hence

$$\angle O_1 = 360° - 90° - 90° - \angle BKC = 180° - \angle BKC = \angle P$$

Similarly, we have

$$\angle O_2 = \angle Q \quad \text{and} \quad \angle O_3 = \angle R$$

It follows that

$$\triangle O_2 O_1 O_3 \sim \triangle ABR \sim \triangle CPB \sim \triangle QCA$$

Done.

Example 4.2.3

(**Napoleon's Triangle**) If three equilateral triangles are constructed outward on the sides of any triangle, centers of these equilateral triangles form an equilateral triangle.

Because all equilateral triangles are similar. The three outward triangles are similar. By the previous conclusion, the triangle formed by their circumcenters is similar to these three triangles which means it is equilateral.

In fact, it can be further shown that if all these three equilateral triangles are drawn inward, their centers also form an equilateral triangle. We will skip its proof here.

4.3 Ptolemy's Theorem

Ptolemy's theorem reveals an important relation among an inscribed quadrilateral's four sides and its two diagonals.

Chapter 4: Circle

> **Theorem 4.3.1   Ptolemy's Theorem**
>
> Given a concyclic quadrilateral $ABCD$, the following relation always holds:
>
> $AB \cdot CD + BC \cdot DA = AC \cdot BD$

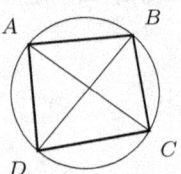

A more general form of the Ptolemy's theorem is expressed as *Ptolemy's inequality* which states that, given any quadrilateral $ABCD$, the following relation always holds:

$$AB \cdot CD + BC \cdot DA \geq AC \cdot BD$$

The equality holds if and only if $ABCD$ is concyclic.

There are several ways to prove Ptolemy theorem. One way is to prove the Ptolemy's inequality first using complex numbers and then show that the equality holds if and only if $ABCD$ is concyclic. We will skip its proof here.

Ptolemy's theorem has two direct corollaries.

> **Property 4.3.1   Corollary 1**
>
> Given a cyclic quadrilateral $ABCD$, if $AB = BC = CA$, then $BD = AD + CD$.

By Ptolemy's theorem:

$DB \cdot AC = DA \cdot BC + DC \cdot AB$

Canceling $AB = BC = CA$ yields the desired claim.

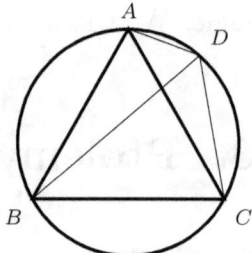

> **Property 4.3.2 Corollary 2**
>
> Given a cyclic quadrilateral $ABCD$, if $\angle B = \angle D = 90°$, then $BD = AC \cdot \sin \angle A$.

$$AC \cdot \sin \angle A$$
$$= AC \cdot \sin(\angle BAC + \angle DAC)$$
$$= AC \cdot \left(\sin \angle BAC \cos \angle DAC + \cos \angle BAC \sin \angle DAC \right)$$
$$= AC \cdot \left(\frac{BC}{AC} \cdot \frac{AD}{AC} + \frac{AB}{AC} \cdot \frac{DC}{AC} \right)$$
$$= (BC \cdot AD + AB \cdot DC)/AC$$
$$= AC \cdot BD / AC$$
$$= BD$$

4.4 Power of a Point

Given a point P and a circle c, the power of point P is defined as the product of distances from P to the two intersection points of any ray emanating from P. P can be outside c, on c, or inside c. The ray can be tangent to c or intersect c at two distinct points.

> **Theorem 4.4.1 Power of a Point Theorem**
>
> The power of a given point P with respect to a given circle c is a constant.

When P is inside c, it is also referred as the *Intersecting Chords Theorem*.
$$PA \cdot PB = PC \cdot PD$$

When P is outside c, it is also referred as the *Intersecting Secants Theorem*. This relation still holds if a secant becomes a tangent line.

Chapter 4: Circle

$$PA \cdot PB = PC \cdot PD = PT^2$$

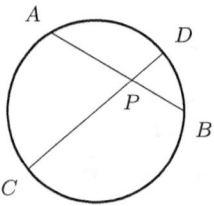
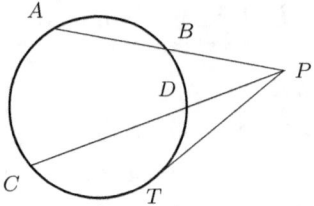

intersecting chords   intersecting secants

When P locates on the circle, it is obvious that the product always equals 0. Thus the equality still holds.

Both intersecting chords theorem and intersecting secants theorem can be proved by using similar triangles. For example, in the former case, the result can be directly derived by noticing $\triangle PAC \sim \triangle PDB$. In the later case, we find $\triangle PAD \sim \triangle PCB$ and $\triangle PCT \sim \triangle PTD$.

They key to apply the power of a point theorem is to construct appropriate chords or scants. Let's look at the following example to understand how.

Example 4.4.1

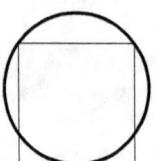
As shown, two vertices of a square are on the circle and one side is tangent to the circle. If the side length of the square is 8, find the radius of the circle.

This problem can be solved in several different ways. For example, one way is to establish several equations by employing Pythagorean theorem and then solve these equations. Having said that, appropriately constructing a chord provides us with a neat alternative solution.

Chapter 4: Circle

Solution

Let the circle's radius be r. Draw a diameter that is perpendicular to the bottom side. Then we have

$$(2r - 8) \times 8 = 4 \times 4 \implies r = 5$$

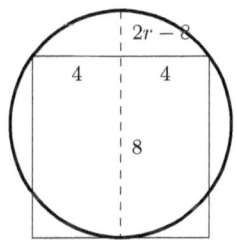

Done.

4.5 Circumscribed Quadrilateral

A circumscribed quadrilateral, sometime referred as a tangential quadrilateral, is a convex quadrilateral whose sides are all tangent to an inside circle.

Theorem 4.5.1 Pitot's Theorem

If $ABCD$ is a circumscribed quadrilateral, then sums of lengths of opposite sides must equal:

$$AB + CD = AC + BD$$

The validity of this theorem is obvious because $AE = AH$, $BE = BF$, $CF = CG$, and $DG = DH$.

Chapter 4: Circle

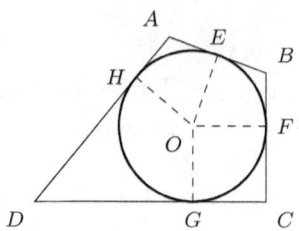

The converse of this theorem also holds. That is, if $AB + CD = AD + BC$, then $ABCD$ is a circumscribed quadrilateral.

Example 4.5.1

If $ABCD$ is a circumscribed quadrilateral, then the four interior angle bisectors meet at the center of its inscribed circle.

Proof

As shown in the diagram above, draw four radii from the incenter O to the four tangent points E, F, G, and H.

Because $\angle OEB = \angle OFB = 90°$ and $OE = OF$, we conclude OB must bisect $\angle B$. Similarly, OC, OD and OA must bisect their corresponding angles. This means the four interior angle bisectors meet at O.

$$QED$$

It can also be shown that AC, BD, EG, and FH are concurrent. This is called *Newton's theorem*. Its proof will be given in *Chapter 5*'s practice.

4.6 Practice

Practice 1

Given a regular pentagon, show that the length ratio between its diagonal and its side equals the golden ratio, $\frac{\sqrt{5}+1}{2}$.

Practice 2

Given two lines tangent to a circle O at points B and C passing a common point A, show that circle O passes through the incenter of $\triangle ABC$.

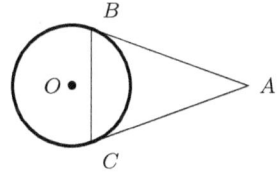

Practice 3

(Brahmagupta Theorem) If the two diagonals of a cyclic quadrilateral is perpendicular, then a line passing the intersection point and perpendicular to one side always bisects the opposite side. Show that $BQ = CQ$, if $BD \perp AC$ and $PM \perp AD$.

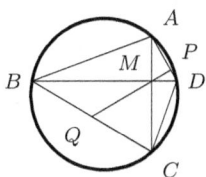

Chapter 4: Circle

Practice 4

(**Carnot's theorem**) Let O be the circumcenter of $\triangle ABC$, $OD \perp BC$, $OD \perp AC$, and $OF \perp AB$ where D, E, and F are the feet on the relevant sides, respectively. If R and r are this triangle's circumradius and inradius, then it always true that $OD + OE + OF = R + r$ when OD, OE, and OF are properly signed. Properly signed means that if the line segment does not pass $\triangle ABC$'s interior, its value should be negative. Otherwise it should be positive.

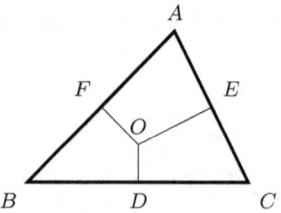

Practice 5

(**Butterfly Theorem**) As shown, in a circle, let chord AB and CD intersect at point M which is the midpoint of chord PQ. If AD meets PQ at point X, and BC meets PQ at point Y. Show M is the midpoint of XY too.

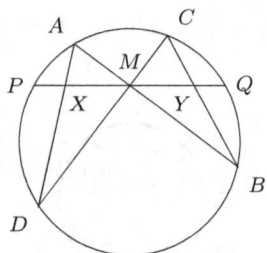

Practice 6

As shown, points X and Y are on the extension of BC in $\triangle ABC$ such that the order of these four points are X, B, C, and Y. Meanwhile, they satisfy the relation $BX \cdot AC = CY \cdot AB$. Let O_1 and O_2 be the circumcenters of $\triangle ACX$ and $\triangle ABY$, respectively. If $O_1 O_2$ intersects AB and AC at U and V, show that $\triangle AUV$ is isosceles.

(Ref 2016 China)

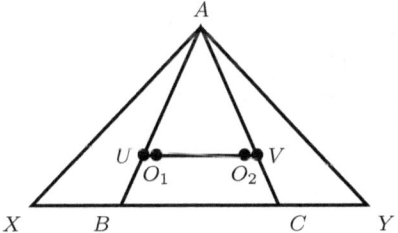

Practice 7

(**Erdös-Mordell Inequality**) Let P be a point inside $\triangle ABC$ and let d_a, d_b, and d_c be the distances from P to BC, CA, and AB, respectively. Prove

$$PA + PB + PC \geq 2(d_a + d_b + d_c)$$

Equality holds if and only if $\triangle ABC$ is equilateral and P is the incenter.

Practice 8

Let P be a point inside $\triangle ABC$. Show that at least one of $\angle PAB$, $\angle PBC$, and $\angle PCA$ is less than or equal to $30°$.

(Ref 1991 IMO)

Practice 9

Let BD be the angle bisector of angle B in $\triangle ABC$ with D on side AC. The circumcircle of $\triangle BDC$ meets AB at E, while the circumcircle of $\triangle ABD$ meets BC at F. Prove that $AE = CF$.

(Ref 1996 St. Petersburg)

Practice 10

(Miquel's Theorem) In $\triangle ABC$, let points A', B', and C' locate on BC, CA, and AB, respectively. Show that the three circumcircles of $\triangle AB'C'$, $\triangle BC''A'$ and $\triangle CA'B'$ are concurrent. Its concurrent point is called the *Miquel point*.

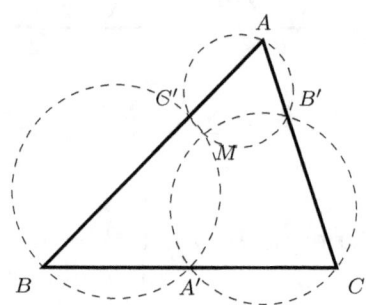

Practice 11

Let $ABCD$ be a circumscribed quadrilateral and O be its incenter. Show that

$$OA \cdot OC + OB \cdot OD = \sqrt{abcd}$$

where a, b, c, and d are lengths of its four sides.

Chapter 5

Cevian and Transversal

5.1 Cevian and Transversal Defined

Cevian and transversal are two important geometry concepts. They are closely related to Ceva's theorem and Menelaus' theorem.

A *cevian* is any line segment in a triangle with one endpoint on a vertex and the other on its opposite side, or its extension.

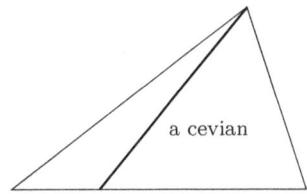

Examples of cevians include median, altitude, angle bisector and so on.

A *transversal* is a straight line that passes two or more lines in the same plane at distinct points.

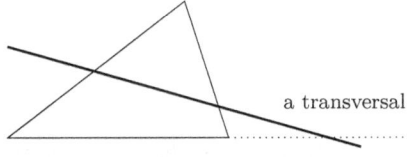

Chapter 5: Cevian and Transversal

Cevians and transversals often appear in triangle related problems which usually can be tackled by Ceva's theorem and Menelaus' theorem.

5.2 Ceva's Theorem

Ceva's theorem is about concurrent cevians. Though, in most cases, the intersecting point locates inside the triangle, the conclusion still holds if it lies outside.

Theorem 5.2.1   Ceva's Theorem

Let P be a point which is not on any of $\triangle ABC$'s sides. Lines AP, BP, and CP intersect opposite sides at D, E, and F. Ceva's theorem states that:

$$\frac{AF}{FB} \cdot \frac{BD}{DC} \cdot \frac{CE}{EA} = 1 \quad (5.1)$$

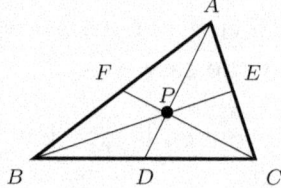

Correctly constructing these segments is critical when employing Ceva's theorem or Menelaus' theorem[1]. The rules to follow can be described as:

- Always starts from one vertex, e.g. A

- Goes to the section point, e.g. F, regardless of whether this point locates on the side itself or the side's extension

- Moves to the next vertex, e.g. B

- Repeat previous steps till coming back to the starting vertex

---

[1]Menelaus' theorem will be discussed in the next section.

Ceva's theorem can be proved in several different ways. One of them is to employ the area method.

$$\frac{AF}{FB} \cdot \frac{BD}{DC} \cdot \frac{CE}{EA} = \frac{S_{\triangle APC}}{S_{\triangle BPC}} \cdot \frac{S_{\triangle APB}}{S_{\triangle APC}} \cdot \frac{S_{\triangle BPC}}{S_{\triangle APB}} = 1$$

Using the area method to solve geometry problems is discussed in the book *Geometry Techniques* by the same author. Several commonly seen patterns, including the one which is utilized in the above proof, will be explained in *Section 6.1.2* on *page 51*.

5.3 Menelaus' Theorem

Length of a segment can be negative. Given a segment, we can assume one direction is positive. Then the sign of its length can be determined by the direction of this segment. So for example, in the following diagram, AB is positive, while BA is negative.

$$A(x_A) \qquad\qquad B(x_B) \qquad +$$

This is consistent with corresponding concepts in coordinate geometry or vectors. Because $AB = x_B - x_A$ and $BA = x_A - x_B$, they must be opposite numbers. Correspondingly, we have

$$AB = -BA \quad \text{or} \quad \frac{AB}{BA} = -1$$

While in many cases it is acceptable to ignore signs of segments' lengths and just to focus on their absolute values, sometimes sign is significant. Typically, Menelaus' theorem is expressed using signed segments.

Chapter 5: Cevian and Transversal

> **Theorem 5.3.1 Menelaus' Theorem**
>
> Given any triangle ABC and a transversal line that crosses BC, CA, and AB, or their extensions at points D, E, and F, respectively. Menelaus' theorem states that:
>
> $$\frac{AF}{FB} \cdot \frac{BD}{DC} \cdot \frac{CE}{EA} = -1$$
>
>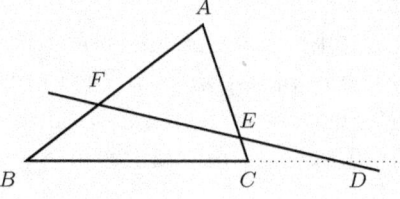

The negative sign of the right side comes from the fact that either one or three intersection points must be on the extensions of corresponding sides which makes either one or three terms on the left side have negative values.

For example, in the diagram above, D locates on the extension of BC. Consequently, BD/DC must be a native value while the other two terms are both positive.

5.4 Trigonometric Ceva's Theorem

While relatively less known, Ceva's theorem can be expressed in trigonometric form. This alternative is described below.

> **Theorem 5.4.1 Ceva's Theorem in Trigonometric Form**
>
> Given $\triangle ABC$ and its three concurrent cevians, AD, BE, and CF, then
>
> $$\frac{\sin \angle CAD}{\sin \angle BAD} \cdot \frac{\sin \angle ABE}{\sin \angle CBE} \cdot \frac{\sin \angle BCF}{\sin \angle ACF} = 1$$
>
>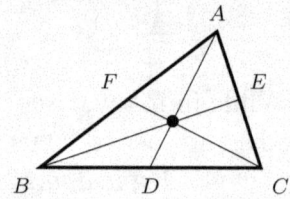

Chapter 5: Cevian and Transversal

5.5 Concurrent Lines

It is important to note that the converses of both Ceva's theorem and Menelaus' theorem also hold. For example, the converse of Ceva's theorem states that if *Equation 5.1* on *page 40* holds, then AD, BE, and CF must be concurrent.

They are often used to prove multiple lines intersect at a single point.

Example 5.5.1

Show that three medians of any triangle are concurrent.

This is a basic conclusion in elementary geometry. There exist several different proofs among which utilizing the converse of Ceva's theorem is one of the simplest.

Proof

As shown, let points D, E, F be the midpoints of BC, CA, AB, respectively. Then we have

$$\frac{AF}{FB} \cdot \frac{BD}{DC} \cdot \frac{CE}{EA} = 1 \times 1 \times 1 = 1$$

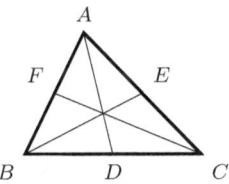

By the converse of Ceva's theorem, we can conclude that AD, BE, and CF are concurrent.

QED

Chapter 5: Cevian and Transversal

5.6 Practice

Practice 1

Show that a triangle's three inner angle bisectors are concurrent.

Practice 2

In $\triangle ABC$, let inner angle bisectors of $\angle B$ and $\angle C$ intersect their opposite sides at E and F, respectively. Let the exterior angle bisector of $\angle A$ intersect BC's extension at D. Show that D, E, and F are collinear.[a]

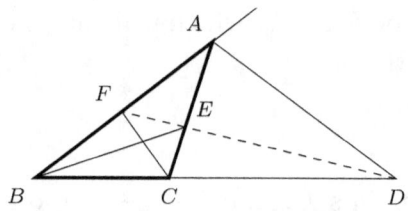

[a] This problem can also be solved by the physics method which is discussed in the book *Geometry Techniques*.

Practice 3

Given $\triangle ABC$, construct three squares outwards using its sides as bases, as shown. Point A_1, B_1, and C_1 are the midpoints of relevant sides. Show that AA_1, BB_1, and CC_1 are concurrent.

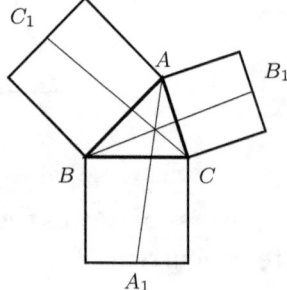

Practice 4

Given $\triangle ABC$, construct three outward similar isosceles triangles using its sides as bases, $\triangle A_1BC$, $\triangle B_1CA$ and $\triangle C_1AB$, respectively. Show that AA_1, BB_1, and CC_1 are concurrent.

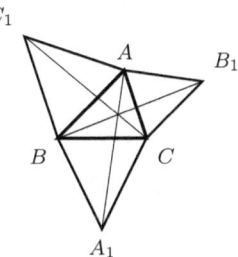

Practice 5

As shown in diagram below, find the degree measure of $\angle ADB$.

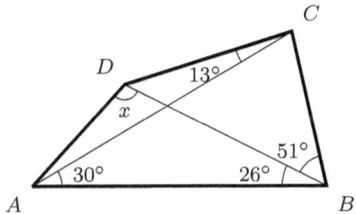

Practice 6

The diagonals AC and CE of a regular hexagon $ABCDEF$ are divided by inner points M and N such that

$$AM : AC = CN : CE = r$$

Determine r if B, M, and N are collinear.[a]

(Ref 1982 IMO)

---

[a]This problem can also be solved by using the coordinate method which is discussed in the book *Geometry Techniques*.

Chapter 5: Cevian and Transversal

Practice 7

(Routh's Theorem) In $\triangle ABC$, let points D, E, and F be on BC, CA, and AB, respectively. If $AF : FB = x$, $BD : DC = y$, and $CE : EA = z$, then

$$S_{\triangle PQR} : S_{\triangle ABC} = \frac{(xyz - 1)^2}{(xy + y + 1)(yz + z + 1)(zx + x + 1)}$$

where P, Q, and R are the intersection points of AD, BE and CF, as shown.

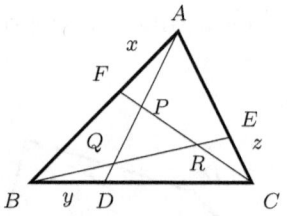

Practice 8

(Pascal's Theorem) Let A, B, C, D, E, and F be points on a circle (not necessarily in cyclic order). Let AB and DE meet at P, BC and EF meet at Q, CD ad FA meet at R. Prove P, Q, and R are collinear.

Practice 9

(**Newton's Theorem**) A circle is inscribed in quadrilateral $ABCD$ with sides touch the circle at E, F, G, and H, as shown. Prove AC, BD, EG, and FH are concurrent.

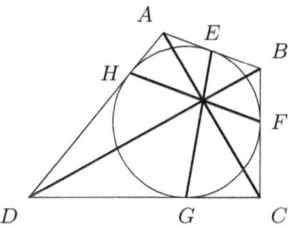

Practice 10

(**Desargues' Theorem**) Given two triangles ABC and $A'B'C'$. Suppose that lines AA', BB', and CC' are concurrent. Let AB and $A'B'$, BC and $B'C'$, CA and $C'A$ intersect at X, Y, and Z, respectively. Show that X, Y, Z are collinear.

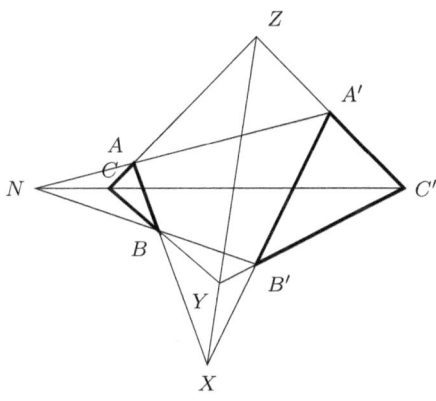

Chapter 5: Cevian and Transversal

Chapter 6

Additional Topics

6.1 Area Computation

This section expands from basic area computation formulas which are taught in school classrooms to some more advanced ones that appear only in math competitions.

6.1.1 Triangle

There are many possible ways to compute a triangle's area under different scenarios. Therefore it is essential for math competition participants to become familiar with all these formulas so they can choose the most appropriate one as needed.

Please note that all the notations used in following formulas are described in *Conventions* on *page 2*.

- Base-Altitude:
$$S = \frac{1}{2} a h_a \tag{6.1}$$

Chapter 6: Additional Topics

- Side-Angle-Side:
$$S = \frac{1}{2}ab\sin C \tag{6.2}$$

- Side-Side-Side (Heron's Formula):
$$S = \sqrt{p(p-a)(p-b)(p-c)} \tag{6.3}$$

- Circumradius - Angles:
$$S = 2R^2 \sin A \sin B \sin C \tag{6.4}$$

- Circumradius - Sides:
$$S = \frac{abc}{4R} \tag{6.5}$$

- Inradius - Semi-perimeter:
$$S = rp \tag{6.6}$$

Both *(6.4)* and *(6.5)* can be derived from *(6.2)* and Law of Sines.

$$S = \frac{1}{2}ab\sin C = \frac{1}{2}(2R\sin A)(2R\sin B)\sin C = 2R^2 \sin A \sin B \sin C$$

$$S = \frac{1}{2}ab\sin C = \frac{1}{2}ab\left(\frac{c}{2R}\right) = \frac{abc}{4R}$$

Combining Heron's formula and *(6.6)* can immediately lead to the following formula which relates a triangle's inradius and its three sides:

$$r = \sqrt{\frac{(p-a)(p-b)(p-c)}{p}} \tag{6.7}$$

Some formulas have various extensions. It is impractical to introduce all of such varieties. Therefore, it is important not only to remember these basic formulas, but also to develop the skill of employing them in a flexible and effective way.

Example 6.1.1

Let point D be on side BC of $\triangle ABC$. Show that

$$S_{\triangle ABC} = \frac{1}{2} \cdot BC \cdot AD \cdot \sin \angle ADB \qquad (6.8)$$

This can be proved by drawing the altitude from vertex A towards base BC. The length of this altitude equals $AD \sin \angle ADB$. Then the to-be-proved result follows by applying the base-altitude formula *(6.1)*.

Alternatively, it can be proved by applying *(6.2)* on $\triangle ABD$ and $\triangle ACD$ respectively and then adding the results together.

It is worthy pointing out that *(6.2)* can be viewed as a special case of *(6.8)* when D coincides with vertex C.

It turns out that *(6.8)* can be further extended to calculate a quadrilateral's area. This is demonstrated in the next example.

Example 6.1.2

Given a quadrilateral, if lengths of its two diagonals are m and n, and one angle formed by these two diagonals is θ, then this quadrilateral's area can be calculated by the following formula:

$$S = \frac{1}{2} \cdot m \cdot n \cdot \sin \theta$$

This result can be proved in a similar way as that used in *Example 6.1.1*.

6.1.2 Ratio of Triangles Areas

In addition to computing triangle's area directly, some problems can be solved by computing area ratio of different triangles. Proof

Chapter 6: Additional Topics

of Ceva's theorem on *page 41* is a good example.

There are several well known basic patterns which are related to area's ratio. They can be abstracted into the following theorem.

> **Theorem 6.1.1   Ratio of Triangle Areas**
>
> If AB and PQ, or their extensions, intersect at point M, then
>
> $$\frac{S_{\triangle PAB}}{S_{\triangle QAB}} = \frac{PM}{QM} \qquad (6.9)$$

Many patterns can be constructed based on this theorem. Some of them are shown below. Proof of Ceva's theorem is a direct application of the 2nd pattern.

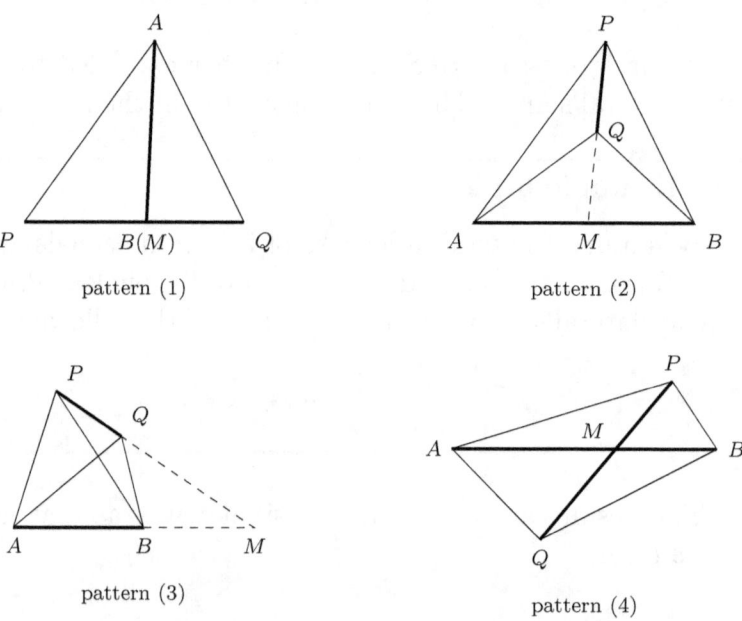

pattern (1)   pattern (2)

pattern (3)   pattern (4)

More varieties and derived patterns can be constructed based on these basic ones. The following example has appeared in some competitions, such as AMC and MathCounts, several times.

Example 6.1.3

Let $ABCD$ be a trapezoid and the areas of those triangles formed by its diagonals be marked as shown. Prove

$S_2 = S4$ and $S_1 \cdot S_3 = S_2 \cdot S_4$

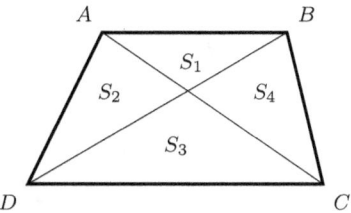

Proof

The first relation holds because

$$AB \parallel CD \implies S_{\triangle ACD} = S_{\triangle BCD} \implies S_2+S_3 = S_4+S_3 \implies S_2 = S_4$$

To prove the 2nd equation, let's denote the intersection of AC and BD as point O. Then

$$S_1 : S_2 = BO : DO \quad \text{and} \quad S_3 : S_4 = DO : BO$$

Multiplying these two equations above yields

$$\frac{S_1}{S_2} \cdot \frac{S_3}{S_4} = 1 \implies S_1 \cdot S_3 = S_2 \cdot S_4$$

$$QED$$

Another useful conclusion which can be established by the area method is given below. Some practice problems will need its conclusion.

Chapter 6: Additional Topics

Example 6.1.4

In $\triangle ABC$, let M be an arbitrary point on BC. A straight line intersects AB, AC and AM at P, Q, and N, respectively. Show that the following relation holds:

$$\frac{AM}{AN} = \frac{AB}{AP} \cdot \frac{CM}{BC} + \frac{AC}{AQ} \cdot \frac{BM}{BC} \qquad (6.10)$$

Proof

Connect PM and QM, as shown.

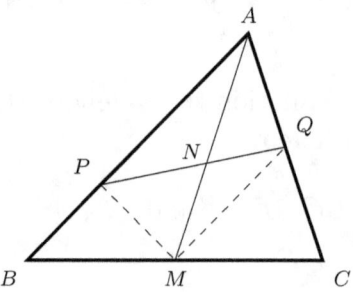

$$\begin{aligned}
\frac{AM}{AN} &= \frac{S_{\triangle APQ} + S_{\triangle MPQ}}{S_{\triangle APQ}} \\
&= \frac{S_{\triangle APM} + S_{\triangle AQM}}{\frac{AP \cdot AQ}{AB \cdot AC} \cdot S_{\triangle ABC}} \\
&= \frac{\frac{AP}{AB} \cdot S_{\triangle ABM} + \frac{AQ}{AC} \cdot S_{\triangle ACM}}{\frac{AP \cdot AQ}{AB \cdot AC} \cdot S_{\triangle ABC}} \\
&= \frac{AC}{AQ} \cdot \frac{BM}{BC} + \frac{AB}{AP} \cdot \frac{CM}{BC} \\
&= \frac{AB}{AP} \cdot \frac{CM}{BC} + \frac{AC}{AQ} \cdot \frac{BM}{BC}
\end{aligned}$$

QED

6.1.3 Quadrilateral

While relatively less known, Heron's formula can be generalized to compute a cyclic quadrilateral's area. Its generalization is called Brahmagupta's formula which has a similar expression as Heron's formula.

> **Theorem 6.1.2 Brahmagupta's formula**
>
> The area S of a cyclic quadrilateral with side lengths of a, b, c, and d is given by
> $$S = \sqrt{(p-a)(p-b)(p-c)(p-d)}$$
> where p is the semi-perimeter of this quadrilateral: $p = \frac{a+b+c+d}{2}$.

Brahmagupta's formula will become Heron's formula when the concyclic quadrilateral becomes a triangle, i.e. $d = 0$.

While Brahmagupta's formula can only compute a cyclic quadrilateral's area, Bretschneider's formula can compute the area of an arbitrary quadrilateral. This formula is given below.

> **Theorem 6.1.3 Bretschneider's formula**
>
> Give a quadrilateral with edges of lengths a, b, c, d in that order and diagonals of lengths m, n, it holds that the area S of this quadrilateral is given by
> $$S = \frac{1}{4}\sqrt{4m^2n^2 - (a^2 + c^2 - b^2 - d^2)^2}$$

From coordinate geometry's perspective, there is a general approach to compute area of a polygon given all its vertices' coordinates. Interested readers may investigate this topic further by referring to appropriate materials.

Chapter 6: Additional Topics

6.1.4 Pick's Theorem

Pick's theorem is a handy formula to solve certain type of area calculation problems. Such problems usually appear in entry level competitions.

> **Theorem 6.1.4   Pick's Theorem**
>
> Given a simple polygon on a grid of equal-distanced points such that all the polygon's vertices are grid points, Pick's theorem states that the polygon's area is given by
>
> $$S = i + \frac{b}{2} - 1$$
>
> where i is the number of grid points inside the polygon and b is the number of grid points on its bound.

In plane geometry, a simple polygon is a polygon whose sides are not intersecting.

Example 6.1.5

Compute the area of the following polygon.

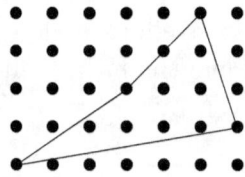

By Pick's theorem, its area equals

$$S = 7 + \frac{5}{2} - 1 = \frac{17}{2}$$

Chapter 6: Additional Topics

6.2 Triangle Centers

A triangle has many special points associated with it. Each has a collection of well known properties. Being familiar with these special points and their associated properties is essential to solve certain types of geometry problems.

As there are many properties, some will be presented as examples and practice problems.

6.2.1 Centroid (Center of Mass)

Centroid is the intersection of three medians. It is often denoted as G. Because of its physics interpretation, centroid is frequently referred as center of mass.

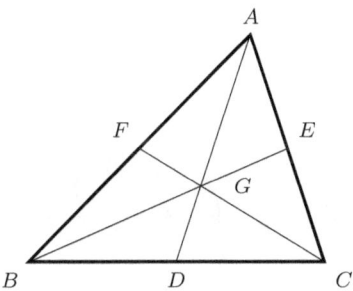

Centroid has many well-known properties. One of its most basic properties is

$$AG : GD = BG : GE = CG : GF = 2 : 1$$

Consequently, the area of $\triangle DEF$ will be $1/4$ of $\triangle ABC$'s area.

Chapter 6: Additional Topics

If coordinates of the three vertices are (x_A, y_A), (x_B, y_B), and (x_C, y_C), then G's coordinates are

$$\left(\frac{x_A + x_B + x_C}{3}, \frac{y_A + y_B + y_C}{3}\right)$$

Example 6.2.1

If G is the centroid of $\triangle ABC$, then

$$BC^2 + 3AG^2 = CA^2 + 3BG^2 = AB^2 + 3CG^2 = \frac{2}{3}(AB^2 + BC^2 + CA^2)$$

Proof

By Apollonius' theorem on *page 83*, we have

$$\left(\frac{3}{2}AG\right)^2 = \frac{1}{2}AB^2 + \frac{1}{2}AC^2 - \frac{1}{4}BC^2$$

Therefore

$$BC^2 + 3AG^2$$
$$= BC^2 + 3 \times \frac{4}{9} \times \left(\frac{1}{2}AB^2 + \frac{1}{2}AC^2 - \frac{1}{4}BC^2\right)$$
$$= \frac{2}{3}(AB^2 + BC^2 + CA^2)$$

QED

6.2.2 Orthocenter

Orthocenter is the intersection of three altitudes. Orthocenter is often denoted as H.

Chapter 6: Additional Topics

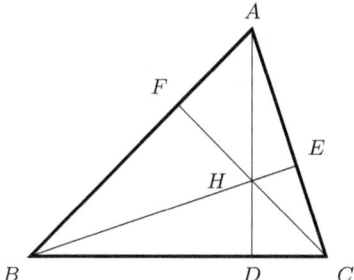

The triangle formed by the three feet, $\triangle DEF$, is called $\triangle ABC$'s orthic triangle. Among all $\triangle ABC$'s inscribed triangles[1], the orthic triangle has the smallest perimeter[2].

Orthocenter related problems often involve concyclic quadrilaterals and similar triangles. Therefore, it is useful to recognize these concyclic quadrilaterals and similar triangles.

There are two basic groups of concyclic quadrilaterals:

- $BCEF$, $CAFD$, and $ABDE$

- $BDHF$, $CDHE$, and $AFHE$

There are several groups of similar triangles. Some of them are listed below

- $\triangle ABC \sim \triangle AEF \sim \triangle DBF \sim \triangle DEC$

- $\triangle AEH \sim \triangle ADC \sim \triangle BEC \sim \triangle BDH$

- $\triangle AFH \sim \triangle ADB \sim \triangle CFB \sim \triangle CDH$

Some additional properties are illustrated in follow examples.

---

[1] An inscribed triangle is such a triangle whose vertices are on three different sides of the original triangle.
[2] This is called *Fagnano's problem*. Its solution is given in the book *Geometry Techniques*.

Chapter 6: Additional Topics

Example 6.2.2

Let H be $\triangle ABC$'s orthocenter. Show that

$$\angle BHC = 180° - \angle A = \angle B + \angle C$$
$$\angle CHA = 180° - \angle B = \angle C + \angle A$$
$$\angle AHB = 180° - \angle C = \angle A + \angle B$$

Without the loss of generality, let's prove the 1st relation above. This can be shown by noticing $AEHF$ are concyclic in the previous diagram.

Example 6.2.3

If H is $\triangle ABC$'s orthocenter, then

$$AB^2 - AC^2 = HB^2 - HC^2$$
$$BC^2 - BA^2 = HC^2 - HA^2$$
$$CA^2 - CB^2 = HA^2 - HB^2$$

These conclusions are natural extensions of *Example 2.3.2* on *page 8*.

Example 6.2.4

Reflection points of the orthocenter with respect to three sides locate on this triangle's circumcircle.

Proof

Extend AH to meet BC at D and circumcircle at D'. Also connect HC and CD'.

Chapter 6: Additional Topics

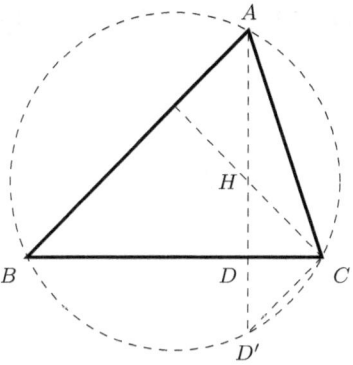

Then

$$\angle BCH = \angle BAD = \angle BCD' \implies \text{Rt}\triangle HCD \cong \text{Rt}\triangle D'CD$$

This means point D' is the reflection point H with respect to BC. The other two reflection points can be proved in the same way.

QED

6.2.3 Circumcenter

The circumcenter is the center of a triangle's circumcircle which is usually denoted as O. It is the intersection of three sides' perpendicular bisectors. The circumcircle's radius is called circumradius which is usually denoted as R.

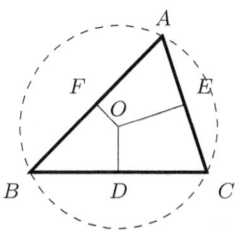

Chapter 6: Additional Topics

Circumradius relates to many theorems and formulas. Some of them are listed below:

i) Law of Sines, *page 15*

ii) Triangle area, *page 50* and so on

iii) Carnot's theorem, *page 94*

iv) Euler's line, *page 131*

In addition, the triangle DEF which is formed by the three feet of perpendicular bisectors is similar to the original triangle ABC with a scaling factor of 0.5. This is because D, E, and F are midpoints of corresponding sides. Therefore DE, EF, and FD must be half in length of corresponding sides.

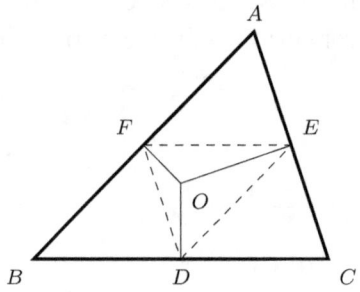

Meanwhile, circumcenter O of $\triangle ABC$ will be the orthocenter of $\triangle DEF$. This is because EF and BC are parallel which implies $OD \perp EF$. Similarly, $OE \perp DF$ and $OF \perp DE$.

6.2.4 Incenter

Incenter is the intersection of three angle bisectors. It is the center of the given triangle's inscribed circle. Incenter is often denoted as I. Correspondingly, the radius of the inscribed circle is called inradius, which is often denoted as r.

Inradius r often relates to the area method. In fact, the general inradius formula which is shown below is derived from Heron's formula.

$$r = \sqrt{\frac{(p-a)(p-b)(p-c)}{p}} \tag{6.11}$$

Example 6.2.5

Let I be the incenter of $\triangle ABC$.
Show that

$$\angle BIC = 90° + \frac{1}{2}\angle A$$

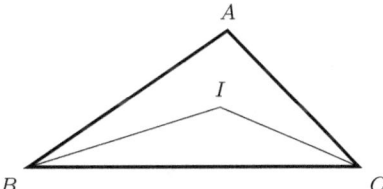

Proof

Because BI and CI bisect $\angle B$ and $\angle C$, respectively, we have

$$\angle BIC = 180° - \frac{1}{2} \times (\angle B + \angle C) = 180° - \frac{1}{2} \times (180° - \angle A) = 90° + \frac{1}{2}\angle A$$

$$QED$$

If the coordinates of $\triangle ABC$'s vertices are (A_x, A_y), (B_x, B_y), and (C_x, C_y), then its incenter's coordinates are given by

$$\left(\frac{aA_x + bB_x + cC_x}{a+b+c}, \frac{aA_y + bB_y + cC_y}{a+b+c} \right)$$

where a, b, and c are corresponding sides' lengths.

Chapter 6: Additional Topics

6.3 Special Points, Lines and Others

Fermat Point

Fermat point P of $\triangle ABC$ is the point that minimize the value of $PA+PB+PC$. If one of the interior angles is at least 120°, that vertex is the Fermat point. Otherwise, it is the unique point that satisfies the condition

$$\angle APB = \angle BPC = \angle CPA = 120°$$

Proving the Fermat point P minimizes the value of $PA+PB+PC$ is discussed in the book *Geometry Techniques*.

Euler's Line

It can be shown that the circumcenter O, centroid G, and orthocenter H are collinear. This line is called Euler's line. In addition, relation $GH = 2GO$ always holds.

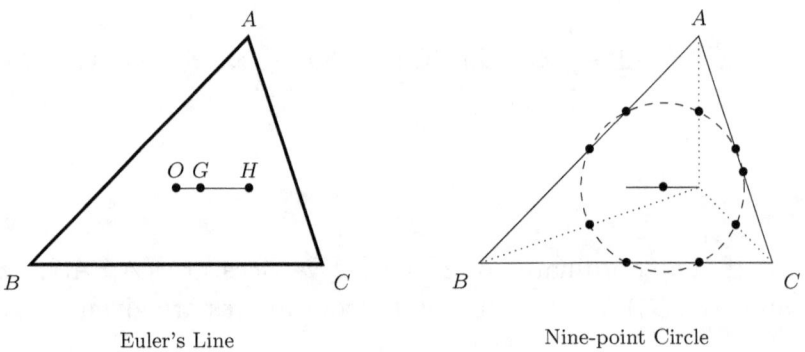

Euler's Line　　　　　　　　Nine-point Circle

Proof of both conclusions is left as a practice.

Nine-point Circle

The nine point circle of a given triangle, shown above right, is the circle that passes midpoint of each side, foot of each altitude, and the midpoint of each vertex and orthocenter. The center of this nine point circle is the midpoint of the Euler's line.

6.4 Practice

Practice 1

Give an outline of proof for the Brahmagupta's formula using Heron's formula and similar triangles.

Practice 2

Derive Brahmagupta's formula (area of a cyclic quadrilateral, *page 55*) from Bretschneider's formula (area of any quadrilateral, *page 55*).

Practice 3

(Subtended Angle Theorem) Let $\angle BAD = \alpha$ and $\angle CAD = \beta$. Show that point D is on BC if and only if the following relation holds:

$$\frac{\sin(\alpha + \beta)}{AD} = \frac{\sin \alpha}{AC} + \frac{\sin \beta}{AB}$$

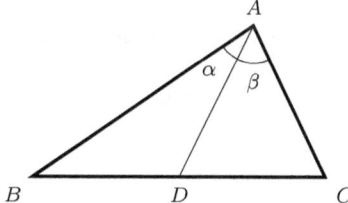

Chapter 6: Additional Topics

Practice 4

(**Steiner's Theorem**) Let M be a point inside $\triangle ABC$ satisfying
$$\frac{\sin \angle BMC}{a} = \frac{\sin \angle CMA}{b} = \frac{\sin \angle CMA}{c}$$
Show that, for any point P inside $\triangle ABC$, the following relation always holds:

$$a \cdot PA + b \cdot PB + c \cdot PC \geq a \cdot MA + b \cdot MB + c \cdot MC$$

The equality holds if and only if M and P coincide.

(Note: In fact, the condition that P is inside $\triangle ABC$ is unnecessary.)

Practice 5

(**Finsler-Hadwiger's Inequalities**) Prove: in any triangle ABC, the following inequalities hold:

$$4\sqrt{3} \cdot S + Q \leq a^2 + b^2 + c^2 \leq 4\sqrt{3} \cdot S + 3Q$$

where S is the triangle's area and $Q = (a-b)^2 + (b-c)^2 + (c-a)^2$.

Tip: By the Schur inequality, we can show that for any positive real number x, y, and z, the following inequality always holds. You may use this to solve this problem.

$$9xzy \geq (x+y+z)(2xy + 2yz + 2zx - x^2 - y^2 - z^2) \tag{6.12}$$

Practice 6

(**Wetizenbock's Inequality**) In any triangle ABC, the following relation holds:

$$a^2 + b^2 + c^2 \geq 4\sqrt{3} \cdot S$$

Practice 7

Does there exist a point P inside $\triangle ABC$ such that any line passing P will divide $\triangle ABC$ into two equal areas?

(Ref 1998 China)

Practice 8

Let G be the centroid of $\triangle ABC$. A line passing through G intersects AB and AC at P and Q, respectively. Show that

$$\frac{AB}{AP} + \frac{AC}{AQ} = 3$$

Practice 9

Let H be the orthocenter of a non-right triangle ABC. A line passing H intersects AB and AC at P and Q, respectively. Show that

$$\frac{AB}{AP} \cdot \tan \angle B + \frac{AC}{AQ} \cdot \tan \angle C = \tan \angle A + \tan \angle B + \tan \angle C$$

Practice 10

(**Japanese Theorem for Cyclic Quadrilaterals**) Let $ABCD$ be a cyclic quadrilateral. Prove that the incenters of $\triangle ABC$, $\triangle BCD$, $\triangle CDA$, and $\triangle DAB$ form a rectangle.

Chapter 6: Additional Topics

Practice 11

Let H be the orthocenter of $\triangle ABC$. Extend AH, BH, CH, and let them meet opposite sides at D, E, and F, respectively. Show that

$$AH \cdot HD = BH \cdot HE = CH \cdot HF$$

Practice 12

Let H be $\triangle ABC$'s orthocenter and R be its circumradius. Show that
$$\frac{AH}{|\cos \angle A|} = \frac{BH}{|\cos \angle B|} = \frac{CH}{|\cos \angle C|} = 2R$$

Practice 13

Given a triangle, show that the distance between an vertex and its orthocenter is twice of the distance between its circumcenter to its opposite side.

Chapter 6: Additional Topics

Practice 14

(**Euler's Line**) Let O, G, and H, be the circumcenter, the centroid, and the orthocenter of $\triangle ABC$, respectively. Show that they are collinear and, in addition, $GH = 2OG$.

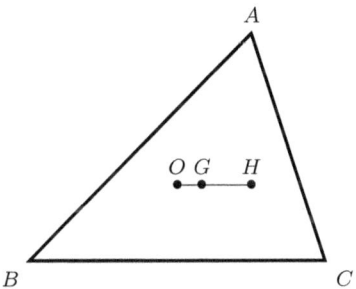

Practice 15

Two mutually tangent congruent circles are internally tangent to a $5-12-13$ triangle, as shown. Find the radius of these two circles.

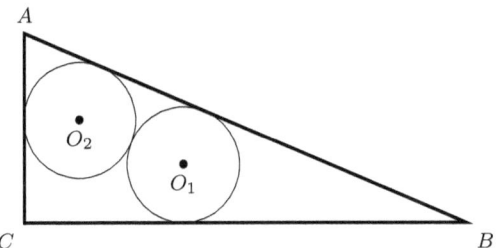

Practice 16

Let G and I be the centroid and incenter of $\triangle ABC$, respectively. If $GI \parallel BC$, show that $AB + AC = 2 \cdot BC$.

Practice 17

Let O be the circumcenter of an acute $\triangle ABC$. If point H lies inside $\triangle ABC$ and satisfies $\angle BAO = \angle HAC$, $\angle ABO = \angle HBC$, show that H is the orthocenter.

Chapter 7

Solutions

Chapter 7: Solutions

7.1 *Chapter 1*

This section is intentionally left blank.

So section numbers of solutions and practices can match.

7.2 Chapter 2

Practice 1

(**Hippocrates Problem**) As shown, three semi-circles are drawn on three sides of right $\triangle ABC$. Show that the sum of shaded areas equals the area of $\triangle ABC$.

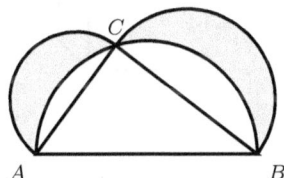

Sum of shaded areas equals

$$\frac{1}{2} \times \left(\frac{1}{2} \cdot AC\right)^2 \pi + \frac{1}{2} \times \left(\frac{1}{2} \cdot BC\right)^2 \pi + \frac{1}{2} \cdot AC \cdot BC - \frac{1}{2} \times \left(\frac{1}{2} \cdot AB\right) \pi$$

Because $AB^2 = AC^2 + BC^2$, the above relation can be simplified to $\frac{1}{2} \cdot AC \cdot BC$ which equals the area of $\triangle ABC$.

Chapter 7: Solutions

Practice 2

Given a rectangular prism with side lengths of 3, 4, and 5, as shown, what is the length of the shortest route from A to C' via surfaces.

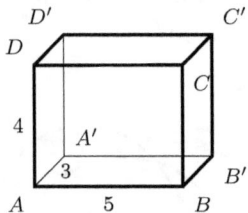

By symmetry, there are two different choices to choose from. One passes edge BC and the other passes CD. In both cases, we can unfold the relevant surfaces to form a flat one and then compute the distance between A and C'.

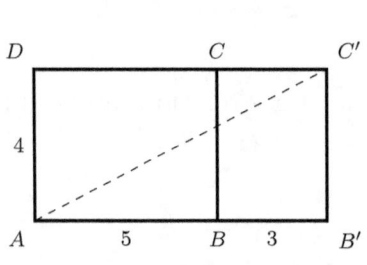
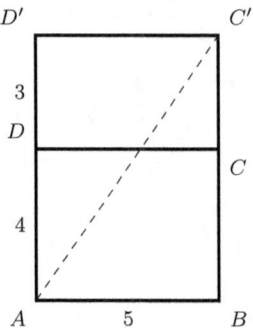

In the first case, $AC' = \sqrt{(5+3)^2 + 4^2} = 4\sqrt{5}$. In the second case, $AC' = \sqrt{5^2 + (4+3)^2} = \sqrt{74}$.

Therefore the final result is

$$min(4\sqrt{5}, \sqrt{74}) = \sqrt{74}$$

Chapter 7: Solutions

Practice 3

(**British Flag Theorem**) Let point P lie inside rectangle $ABCD$. Draw four squares using each of AP, BP, CP, and DP as one side. Show that

$$S_{AA_1A_2P} + S_{CC_1C_2P} = S_{BB_1B_2P} + S_{DD_1D_2P}$$

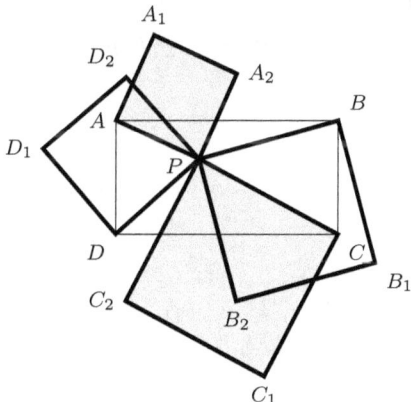

Draw two lines passing through point P such that one is parallel to side AB and the other is parallel to side AD. Let they meet the sides at points M, N, K, and L, as shown below.

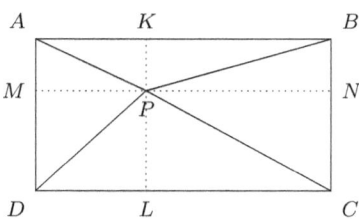

Chapter 7: Solutions

Therefore:

$$\begin{aligned}
S_{AA_1A_2P} &= AP^2 = PK^2 + AK^2 = PK^2 + PM^2 \\
S_{BB_1B_2P} &= BP^2 = PK^2 + BK^2 = PK^2 + PN^2 \\
S_{CC_1C_2P} &= CP^2 = PL^2 + LC^2 = PL^2 + PN^2 \\
S_{DD_1D_2P} &= DP^2 = PL^2 + LD^2 = PL^2 + PM^2
\end{aligned}$$

$$\therefore\ S_{AA_1A_2P} + S_{CC_1C_2P} = PK^2 + PL^2 + PM^2 + PN^2$$
$$S_{BB_1B_2P} + S_{DD_1D_2P} = PK^2 + PL^2 + PM^2 + PN^2$$
$$\implies S_{AA_1A_2P} + S_{CC_1C_2P} = S_{BB_1B_2P} + S_{DD_1D_2P}$$

Practice 4

(**De Gua's Theorem**) In a trirectangular tetrahedron $ABCD$ where A is the shared right-angle corner. Show that

$$S^2_{\triangle BCD} = S^2_{\triangle ABC} + S^2_{\triangle ACD} + S^2_{\triangle ADB}$$

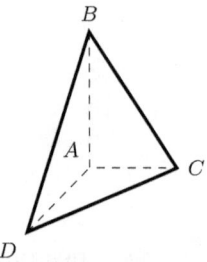

For convenience, let's mark the lengths of three legs as b, c, and d. This is shown in the diagram below.

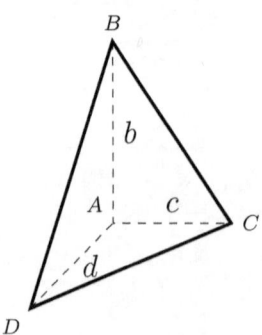

Chapter 7: Solutions

Then

$$S^2_{\triangle ABC} + S^2_{\triangle ACD} + S^2_{\triangle ADB} = \frac{1}{4} \times \left(b^2c^2 + c^2d^2 + d^2b^2\right) \quad (7.1)$$

Meanwhile, the lengths of the base $\triangle BCD$ are $\sqrt{b^2+c^2}$, $\sqrt{c^2+d^2}$, and $\sqrt{d^2+b^2}$. Its area $S_{\triangle BCD}$ can be computed using Heron's Formula[1] directly. Alternatively, the area/volume method[2] also provides a convenient solution.

The distance from point A to the base BCD equals

$$h = \frac{1}{\sqrt{\frac{1}{b^2} + \frac{1}{c^2} + \frac{1}{d^2}}} = \frac{bcd}{\sqrt{b^2c^2 + c^2d^2 + d^2b^2}}$$

Therefore

$$V_{ABCD} = \frac{1}{6} \times bcd = \frac{1}{3} \times S_{\triangle BCD} \times h$$

$$\implies S_{\triangle BCD} = \frac{1}{2}\sqrt{b^2c^2 + c^2d^2 + d^2b^2} \quad (7.2)$$

Comparing (7.1) and (7.2) leads to the desired result.

Practice 5

Let CD be the altitude in right $\triangle ABC$ from the right angle C. If inradii of $\triangle ABC$, $\triangle ACD$, and $\triangle BCD$ are r_1, r_2, and r_3, respectively, show that

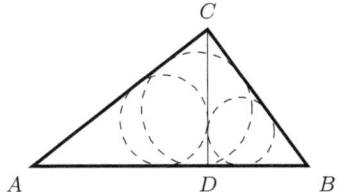

$$r_1 + r_2 + r_3 = CD$$

---

[1] Heron's formula will be discussed in *Chapter 6*.
[2] A detailed discussion of the area/volume method can be found in the book *Geometry Techniques* by the same author.

Let $AC = b$, $BC = a$, and $AB = c$. Then $CD = \frac{ab}{c}$, $AD = \frac{b^2}{c}$, and $BD = \frac{a^2}{c}$ by geometric mean theorem. It follows that

$$\begin{cases} r_1 = (a+b-c)/2 \\ r_2 = (\frac{ab}{c} + \frac{b^2}{c} - b)/2 \\ r_3 = (\frac{ab}{c} + \frac{a^2}{c} - a)/2 \end{cases}$$

Therefore

$$r_1 + r_2 + r_3$$
$$= (a+b-c)/2 + \left(\frac{ab}{c} + \frac{b^2}{c} - b\right)/2 + \left(\frac{ab}{c} + \frac{a^2}{c} - a\right)/2$$
$$= \frac{1}{2} \times \left(\frac{a^2 + b^2 + 2ab}{c} - c\right)$$
$$= \frac{1}{2} \times \left(\frac{c^2 + 2ab}{c} - c\right)$$
$$= \frac{ab}{c}$$
$$= CD$$

Practice 6

Let $\triangle ABC$ be a right triangle where $\angle C = 90°$. Show that if point D is on side BC,

$$AB^2 = DB^2 + DA^2 + 2 \cdot DB \cdot DC$$

If D locates on BC's extension, then

$$AB^2 = DB^2 + DA^2 - 2 \cdot DB \cdot DC$$

Let's first prove the case when D locates on BC.

Chapter 7: Solutions

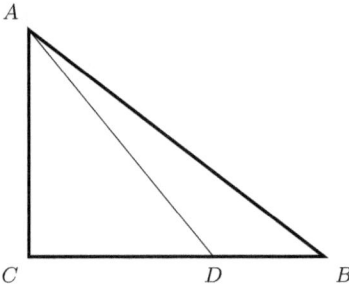

By Pythagorean theorem, we have

$$AC^2 = AB^2 - BC^2 \quad \text{and} \quad AC^2 = AD^2 - DC^2$$

$$\therefore \quad AB^2 - BC^2 = AD^2 - DC^2$$

It follows that

$$\begin{aligned} AB^2 &= AD^2 - DC^2 + BC^2 \\ &= AD^2 - DC^2 + (DC + DB)^2 \\ &= AD^2 - DC^2 + DC^2 + DB^2 + 2 \cdot BD \cdot DC \\ &= DB^2 + DA^2 + 2 \cdot DB \cdot DC \end{aligned}$$

When D locates on the extension of BC, the 2nd step will have $BC = DC - DB$ which will result in a $(-2 \cdot DB \cdot DC)$ term.

Practice 7

Three circles are tangent to each other and also a common line, as shown. Let the radii of circles O_1, O_2, and O_3 be r_1, r_2, and r_3, respectively. Show that

$$\frac{1}{\sqrt{r_3}} = \frac{1}{\sqrt{r_1}} + \frac{1}{\sqrt{r_2}}$$

Chapter 7: Solutions

Connect O_1O_2, $O_1A \perp AB$, $O_2B \perp AB$, and $O_2D \perp O_1A$.

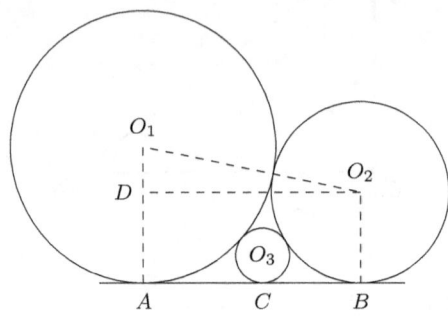

Let's study right $\triangle O_1O_2D$. It is clear that $O_1O_2 = r_1 + r_2$ and $O_1D = r_1 - r_2$. Therefore

$$AB = O_2D = \sqrt{(r_1+r_2)^2 - (r_1-r_2)^2} = 2\sqrt{r_1r_2}$$

Similarly, we can derive

$$AC = 2\sqrt{r_1r_3} \quad \text{and} \quad BC = 2\sqrt{r_2r_3}$$

Because $AB = AC + BC$, we have

$$2\sqrt{r_1r_2} = 2\sqrt{r_1r_3} + 2\sqrt{r_2r_3} \implies \frac{1}{\sqrt{r_3}} = \frac{1}{\sqrt{r_1}} + \frac{1}{\sqrt{r_2}}$$

Practice 8

Let point P be inside an equilateral $\triangle ABC$ such that $AP = 3$, $BP = 4$, and $CP = 5$. Find the side length of $\triangle ABC$.

Because $3, 4$, and 5 form a Pythagorean triplet. Let's try to construct a right triangle by the rotation method[3]. In order to

---
[3] The rotation method is discussed in the book *Geometry Techniques*.

achieve this, rotate $\triangle ABP$ around point A for $60°$ counterclockwise. Because $\triangle ABC$ is equilateral, point B will move to point C. Let the new location of point P be P'. See the diagram (i) below.

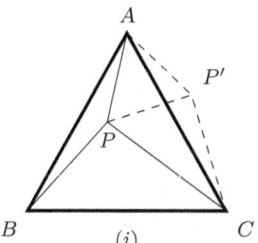 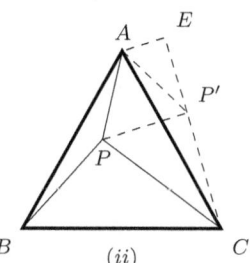

Now we find $AP' = AP = 3$ and $\angle PAP' = 60°$, therefore $\triangle APP'$ is equilateral. This implies $PP' = 3$ too. Obviously, $CP' = BP = 4$ and $CP = 5$ Therefore CPP' is a right triangle by the converse of Pythagorean theorem, or $\angle CP'P = 90°$.

Next, draw $AE \perp PP'$ and let the foot be E. See the diagram (ii) above.

$$\angle AP'E = 180° - \angle PP'C - \angle AP'P = 180° - 90° - 60° = 30°$$

Therefore

$$AE = \frac{1}{2}AP' = \frac{3}{2} \quad \text{and} \quad CE = \frac{\sqrt{3}}{2}AE + CP' = \frac{3\sqrt{3}}{2} + 4$$

Applying Pythagorean theorem on $\triangle AEC$ yields:

$$AC = \sqrt{\left(\frac{3}{2}\right)^2 + \left(\frac{3\sqrt{3}}{2} + 4\right)^2} = \sqrt{25 + 12\sqrt{3}}$$

Practice 9

Let M be a point inside $\triangle ABC$. Draw $MA' \perp BC$, $MB' \perp CA$, and $MC' \perp AB$ such that $BA' = BC'$ and $CA' = CB'$. Prove $AB' = AC'$.

(Ref 1979 China)

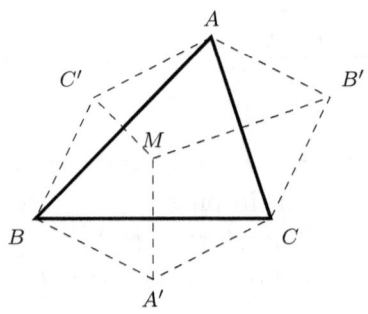

By *Example 2.3.2* on *page 8*:

$$MA' \perp BC \implies A'B^2 - MB^2 = A'C^2 - MC^2$$

$$MB' \perp CA \implies B'C^2 - MC^2 = B'A^2 - MA^2$$

$$MC' \perp AB \implies C'A^2 - MA^2 = C'B^2 - MB^2$$

Adding these three equations together and noticing $BA' = BC'$ and $CA' = CB'$ lead to the desired conclusion.

Chapter 7: Solutions

7.3 Chapter 3

Practice 1

Given a triangle, show that the angle formed by the angle bisector and the altitude from the same vertex equals half of the difference between the two base angles.

In $\triangle ABC$, let AT be the angle bisector and AH be the altitude.

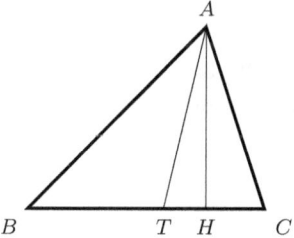

$$\begin{aligned}
2\angle TAH &= (\angle BAH - \angle BAT) + (\angle TAC - \angle HAC) \\
&= \angle BAH - \angle HAC \\
&= (90° - \angle B) - (90° - \angle C) \\
&= \angle C - \angle B
\end{aligned}$$

Practice 2

(Apollonius' Theorem) Let AD be one median of $\triangle ABC$ where point D lies on side BC. Show that the following relation holds:
$$AB^2 + AC^2 = 2 \times (AD^2 + BD^2)$$

Chapter 7: Solutions

By Stewart theorem, we have

$$AB^2 \cdot CD + AC^2 \cdot BD = BC \cdot (AD^2 + BD \cdot CD)$$

Setting $BD = CD$ (because AD is a median) and $BC = BD + CD$ to the above relation will lead to the to-be-proved claim.

Practice 3

Show that when $\triangle ABC$ is a right triangle, Apollonius' theorem (see the previous practice problem) will reduce to Pythagorean theorem.

Suppose $\angle C = 90°$ and median CD meets AB at point C. By the properties of a right triangle, we have $AD = BD = CD = \frac{1}{2}AB$. Therefore

$$BC^2 + AC^2 = 2 \times (CD^2 + AD^2) = 2 \times \left(\frac{1}{4}AB^2 + \frac{1}{4}AB^2\right) = AB^2$$

Practice 4

In a right triangle whose legs are a and b, and hypotenuse is c, two segments drawn from the right angle divide the hypotenuse into three equal parts of length x. If the lengths of these two segments are p and q, show that $p^2 + q^2 = 5x^2$.

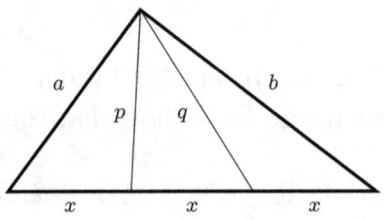

Applying the Apollonius' theorem (or Steward theorem) first on

p and the q yields:

$$\begin{cases} a^2 + q^2 = 2(p^2 + x^2) \\ p^2 + b^2 = 2(q^2 + x^2) \end{cases}$$

Adding these two equations yields

$$a^2 + b^2 + p^2 + q^2 = 2(p^2 + q^2) + 4x^2$$
$$(3x)^2 = p^2 + q^2 + 4x^2$$
$$p^2 + q^2 = 5x^2$$

Practice 5

Given a parallelogram $ABCD$, show that

$$AB^2 + BC^2 + CD^2 + DA^2 = AC^2 + BD^2$$

Let the two diagonals intersect at point O. Then AC and BD mutually bisect each other.

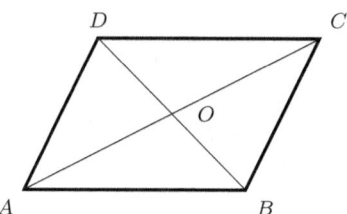

Apply Apollonius' theorem on $\triangle DAC$ leads to:

$$DA^2 + DC^2 = 2 \times (DO^2 + AO^2) = \frac{1}{2} \times (BD^2 + AC^2)$$

Similarly, we shall have:

$$BA^2 + BC^2 = 2 \times (BO^2 + AO^2) = \frac{1}{2} \times (BD^2 + AC^2)$$

Chapter 7: Solutions

Adding them together leads to the desired result.

Practice 6

(**Law of Tangents**) In any triangle, show that

$$\frac{a+b}{a-b} = \frac{\tan(\frac{A+B}{2})}{\tan(\frac{A-B}{2})}$$

By Law of Sines, we have $a = 2R\sin A$ and $b = 2R\sin B$. It follows that

$$a + b = 2R(\sin A + \sin B) = 4R\sin\left(\frac{A+B}{2}\right)\cos\left(\frac{A-B}{2}\right)$$

$$a - b = 2R(\sin A - \sin B) = 4R\cos\left(\frac{A+B}{2}\right)\sin\left(\frac{A-B}{2}\right)$$

$$\therefore \frac{a+b}{a-b} = \frac{\sin\left(\frac{A+B}{2}\right)\cos\left(\frac{A-B}{2}\right)}{\cos\left(\frac{A+B}{2}\right)\sin\left(\frac{A-B}{2}\right)} = \frac{\tan(\frac{A+B}{2})}{\tan(\frac{A-B}{2})}$$

Practice 7

Find the measure of the largest interior angle of $\triangle ABC$ if its side lengths are 3, 5, and 7.

The largest interior angle must be the one opposite to the side with length 7. Without loss of generality, assume it is $\angle C$. By Law of Cosines, we have

$$\cos \angle C = \frac{3^2 + 5^2 - 7^2}{2 \times 3 \times 5} = -\frac{1}{2} \implies \angle C = \boxed{120°}$$

🛈 *Tip: A $3-5-7$ triangle is special because of this property.*

Chapter 7: Solutions

Practice 8

If circle O has an inscribed pentagon whose sides lengths are 3, 3, 5, 5 and 7, in that order. Find the area of this circle.

First, we notice that switching the order of the last two sides does not change the measurement of this circle. This is because by switching $CD = 5$ and $DE = 7$ with $CD' = 7$ and $D'E = 5$, the two triangles $\triangle CDE$ and $\triangle CD'E$ are symmetrical, as shown in the diagram below.

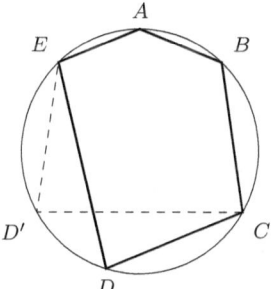 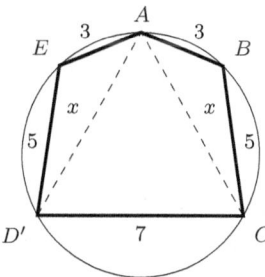

Now let's connect AD' and AC in the side-switched pentagon $ABCD'E$ as shown on the right above. By symmetry, AC must equal AD'. Let their lengths be x.

By the previous practice, we can make an educated guess that $x = 7$. This will lead to $\angle B = \angle E = 120°$. Meanwhile $\angle ACD' = \angle AD'C = 60°$. This will satisfy the condition that $ABCD'E$ is inscribed because both $ABCD'$ and $ACD'E$ are inscribed (the sum of opposite interior angles is 180°).

This educated guess can be verified by applying Law of Cosines on $\triangle ABC$ and $\triangle ACD'$, respectively.

$$\begin{cases} \cos \angle ABC = \frac{3^2+5^2-x^2}{2\times 3\times 5} \\ \cos \angle AD'C = \frac{x^2+7^2-x^2}{2\times 7\times x} \end{cases}$$

Chapter 7: Solutions

Notice that
$$\angle ABC + \angle AD'C = 180° \implies \cos \angle ABC = -\cos \angle AD'C$$

We have
$$\frac{3^2 + 5^2 - x^2}{2 \times 3 \times 5} = -\frac{x^2 + 7^2 - x^2}{2 \times 7 \times x}$$

This equation has only one real root $x = 7$.

It follows that the radius of this circle is $7 \times \frac{\sqrt{3}}{2} \times \frac{2}{3} = \frac{7}{\sqrt{3}}$.
Therefore the area of this circle equals $\boxed{\dfrac{49\pi}{3}}$.

Practice 9

Given $\triangle ABC$, $\angle C$ is quadrisected (divided into four equal angles) by the altitude, the angle bisector, and the median from that vertex C. Find the measurement of $\angle C$.

Without loss of generality, let's assume $BC = 1$.

Let $BH = x$. Then in $\triangle CIB$, because $\angle BCH = \angle ICH$ and $CH \perp BI$, we have $HI = x$ and $CI = 1$.

Let $MI = y$. Because CM is a median, we find $AM = 2x + y$.

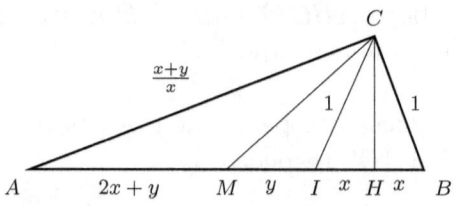

Applying the angle bisector theorem on $\triangle ABC$:
$$\frac{AC}{AI} = \frac{BC}{BI} \implies AC = \frac{x+y}{x}$$

Applying the angle bisector theorem on $\triangle CAI$:

$$\frac{CA}{AM} = \frac{CI}{IM} \implies \frac{\frac{x+y}{x}}{2x+y} = \frac{1}{y} \implies \frac{y}{x} = \sqrt{2}$$

Now apply the angle bisector theorem on $\triangle CMH$ again:

$$\frac{CM}{CH} = \frac{MI}{HI} = \sqrt{2}$$

Because this is a right triangle, it must hold that $\angle MCH = 45°$. Therefore we conclude

$$\angle ACB = 2\angle MCH = \boxed{90°}$$

Practice 10

(Law of Cosines in 3-Dimensional Space) In tetrahedron $ABCD$, let the areas of $\triangle BCD$, $\triangle ACD$, $\triangle ABD$, and $\triangle BCD$ be a, b, c, and d, respectively. Also let (a,b), (b,c), and (c,a) be the angles between faces $DBC - DCA$, $DCA - DAB$, and $DAB - DBC$, respectively. Prove

$$d^2 = a^2 + b^2 + c^2 - 2ab\cos(a,b) - 2bc\cos(b,c) - 2ca\cos(c,a)$$

Using the projection method[4], we find:

$$d = a\cos(a,d) + b\cos(b,d) + c\cos(cd) \tag{7.3}$$

where angles (a,d), (b,d), and (c,d) are similarly defined.

By symmetry, the following relations must hold too:

$$\begin{cases} a = b\cos(b,a) + c\cos(c,a) + d\cos(d,a) \\ b = c\cos(c,b) + d\cos(d,b) + a\cos(a,b) \\ c = d\cos(d,c) + a\cos(a,c) + b\cos(b,c) \end{cases}$$

---
[4]The projection method is discussed in the book *Geometry Techniques*.

Chapter 7: Solutions

Rearranging the three above equations leads to

$$\begin{cases} \cos(d,a) = \frac{1}{d}\Big(a - b\cos(b,a) - c\cos(c,a)\Big) \\ \cos(d,b) = \frac{1}{d}\Big(b - c\cos(c,b) - a\cos(a,b)\Big) \\ \cos(d,c) = \frac{1}{d}\Big(c - a\cos(a,c) - b\cos(b,c)\Big) \end{cases}$$

Setting these into *(7.3)*:

$$\begin{aligned} d = &\ \frac{a}{d}\Big(a - b\cos(b,a) - c\cos(c,a)\Big) \\ &+ \frac{b}{d}\Big(b - c\cos(c,b) - a\cos(a,b)\Big) \\ &+ \frac{c}{d}\Big(c - a\cos(a,c) - b\cos(b,c)\Big) \end{aligned}$$

Multiplying both sides by d and consolidating similar terms yields the desired result.

7.4 Chapter 4

Practice 1

Given a regular pentagon, show that the length ratio between its diagonal and its side equals the golden ratio, $\frac{\sqrt{5}+1}{2}$.

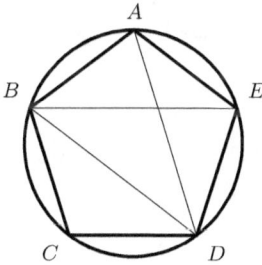

Let the length of a diagonal be b and a side be a. Applying Ptolemy's theorem to $ABDE$ yields:

$$BE \cdot AD = AE \cdot BD + AB \cdot DE \implies b^2 = ab + a^2$$

Because $a \neq 0$, we can divide both sides by a^2 and rearrange to:

$$\left(\frac{b}{a}\right)^2 - \left(\frac{b}{a}\right) - 1 = 0$$

This equation has only one positive root:

$$\frac{b}{a} = \frac{1+\sqrt{5}}{2}$$

Chapter 7: Solutions

Practice 2

Given two lines tangent to a circle O at points B and C passing a common point A, show that circle O passes through the incenter of $\triangle ABC$.

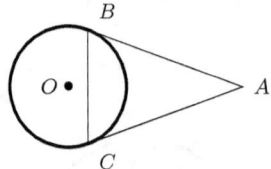

Let the angle bisectors of $\angle B$ and $\angle C$ intersect at point Q. Then Q is $\triangle ABC$'s incenter. We are going to show that Q locates on circle O.

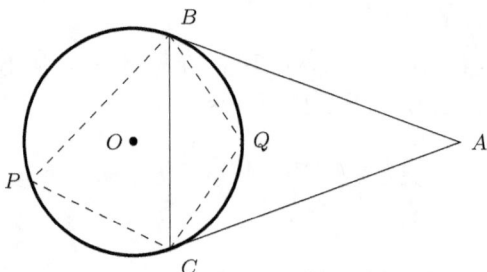

Let P be any point on circle O. Then because AB and AC are tagent lines, we have

$$\angle ABC = \angle ACB = \angle P \implies \angle ABC + \angle ACB = 2\angle P$$

Because BQ and CQ are angle bisectors, we have

$$\angle QBC + \angle QCB = \angle P$$

This implies

$$\angle Q + \angle P = \angle Q + (\angle QBC + \angle QCB) = 180°$$

which means $PBQC$ are concyclic, or point Q is on circle O.

Practice 3

(Brahmagupta Theorem) If the two diagonals of a cyclic quadrilateral is perpendicular, then a line passing the intersection point and perpendicular to one side always bisects the opposite side. Show that $BQ = CQ$, if $BD \perp AC$ and $PM \perp AD$.

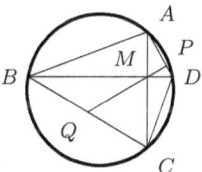

We can first show that $MQ = BQ$ by proving $\angle MBQ = \angle BMQ$.

Because MP is an altutide of right triangle AMD, it must hold that $\angle DAM = \angle DMP$. Because $ABCD$ is cyclic, $\angle DAM = \angle DBQ$. Meanwhile $\angle DMP = \angle BMQ$. Hence, we find

$$\angle MBQ = \angle BMQ \implies MQ = BQ$$

Similarly, we can show that

$$\angle QMC = \angle QCM \implies QM = QC$$

Therefore point Q must be the midpoint of BC.

Chapter 7: Solutions

Practice 4

(**Carnot's theorem**) Let O be the circumcenter of $\triangle ABC$, $OD \perp BC$, $OD \perp AC$, and $OF \perp AB$ where D, E, and F are the feet on the relevant sides, respectively. If R and r are this triangle's circumradius and inradius, then it always true that $OD + OE + OF = R + r$ when OD, OE, and OF are properly signed. Properly signed means that if the line segment does not pass $\triangle ABC$'s interior, its value should be negative. Otherwise it should be positive.

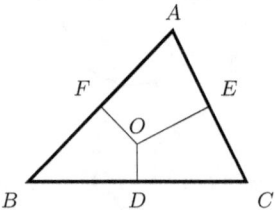

Denote $a = BC$, $b = AC$, and $c = AB$. We discuss three possible scenarios.

Case 1: $\triangle ABC$ is a right triangle where $\angle C = 90°$.

In this case, O is the the midpoint of AB. $OD = \frac{b}{2}$, $OE = \frac{a}{2}$, $OF = 0$, $R = \frac{c}{2}$, and $r = \frac{a+b-c}{2}$. Hence it is obviously holds that

$$OD + OE + OF = R + r = \frac{a+b}{2}$$

Case 2: $\triangle ABC$ is an acute triangle.

In this case, O locates inside the triangle. Therefore all of OD, OE and OF are positive.

Let's connect AO and EF. Because O is the circumcenter, D, E, and F must be the midpoints of corresponding sides. It follows that $EF = \frac{BC}{2} = \frac{a}{2}$.

Chapter 7: Solutions

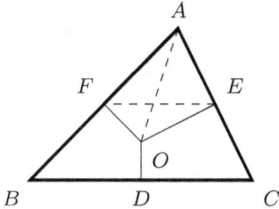

Because $OE \perp AC$ and $OF \perp AB$, we can conclude A, E, O, and F are concyclic. By Ptolemy's theorem:

$$AF \cdot OE + AE \cdot OF = AO \cdot EF \implies \frac{c}{2} \cdot OE + \frac{b}{2} \cdot OF = R \cdot \frac{a}{2}$$

or

$$c \cdot OE + b \cdot OF = a \cdot R \qquad (7.4)$$

Similarly, we can derive

$$a \cdot OF + c \cdot OD = b \cdot R \qquad (7.5)$$

$$b \cdot OD + a \cdot OE = c \cdot R \qquad (7.6)$$

Meanwhile, we have

$$a \cdot OD + b \cdot OE + c \cdot OF = 2 \cdot S_{\triangle ABC} \qquad (7.7)$$

Adding *(7.4)*, *(7.5)*, *(7.6)*, and *(7.7)* leads to

$$(a+b+c)(OD+OE+OF) = (a+b+c)R + 2 \cdot S_{\triangle ABC} \qquad (7.8)$$

Using the area method, setting $2 \cdot S = r \cdot (a+b+c)$ to the above equation, and canceling common factor $(a+b+c)$ lead to the desired result.

Case 3: $\triangle ABC$ is an obtuse triangle where $\angle A > 90°$.

Without loss of generality, let's assume $\angle A > 90°$. As a result, $OD < 0$. Connect OB and OC. By a similar argument, it can be shown that *(7.8)* still holds. This is because when DO is negative, the area of ABC equals the difference between the area of $ABOC$ and BOC.

Chapter 7: Solutions

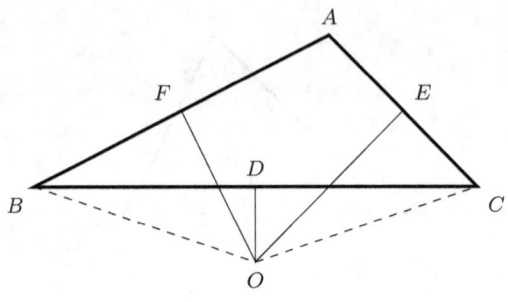

Practice 5

(**Butterfly Theorem**) As shown, in a circle, let chord AB and CD intersect at point M which is the midpoint of chord PQ. If AD meets PQ at point X, and BC meets PQ at point Y. Show M is the midpoint of XY too.

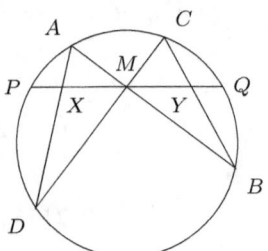

Suppose U and V are the midpoints of AD and BC, respectively. Let's connect them with the center of circle O. In addition, also connect XO, YO, MU, and MV.

Chapter 7: Solutions

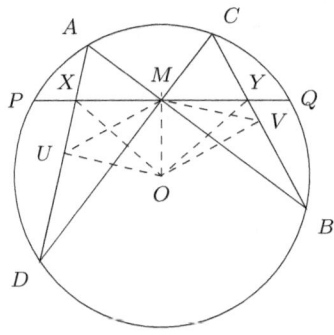

Because $\angle A = \angle C$ and $\angle D = \angle B$, we find $\triangle ADM \sim \triangle CBM$. Because U and V are midpoints of AD and CB, respectively, it must hold that $\triangle AUM \sim \triangle CVM$ which implies

$$\angle AUM = \angle CVM \tag{7.9}$$

Because U is the midpoint of AD, it must hold that $OU \perp AD$. Similarly, $OM \perp PQ$. Therefore we find O, M, X, and U are concyclic. Consequently, it must hold that $\angle AUM = \angle XOM$. By the same reasoning, we can conclude $\angle CVM = \angle YOM$.

By *(7.9)*, we find $\angle XOM = \angle YOM$. Noticing that $OM \perp XY$, this leads to the conclusion that $XM = YM$.

Practice 6

As shown, points X and Y are on the extension of BC in $\triangle ABC$ such that the order of these four points are X, B, C, and Y. Meanwhile, they satisfy the relation $BX \cdot AC = CY \cdot AB$. Let O_1 and O_2 be the circumcenters of $\triangle ACX$ and $\triangle ABY$, respectively. If O_1O_2 intersects AB and AC at U and V, show that $\triangle AUV$ is isosceles.

(Ref 2016 China)

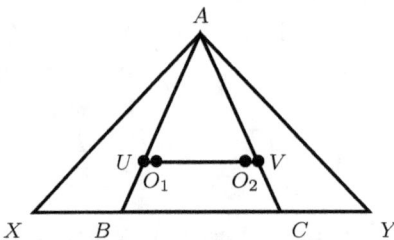

Let AZ bisects $\angle BAC$ and meet BC at Z.

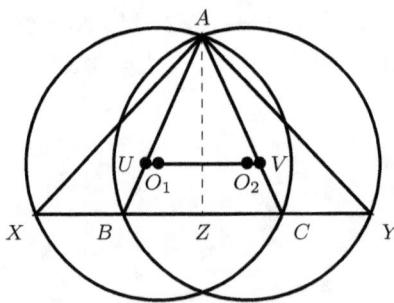

We have

$$\frac{BZ}{CZ} = \frac{AB}{AC} = \frac{BX}{CY} = \frac{BZ + BX}{CZ + CY} = \frac{XZ}{YZ}$$

or

$$BZ \cdot ZY = CZ \cdot ZX$$

Chapter 7: Solutions

This implies that Z lies on this two circle's radical axis. It follows that $AZ \perp BC$ and $AZ \perp UV$. Because AZ also bisects $\angle UAV$, it must hold that $AU = AV$.

Practice 7

(**Erdös-Mordell Inequality**) Let P be a point inside $\triangle ABC$ and let d_a, d_b, and d_c be the distances from P to BC, CA, and AB, respectively. Prove

$$PA + PB + PC \geq 2(d_a + d_b + d_c)$$

Equality holds if and only if $\triangle ABC$ is equilateral and P is the incenter.

Let X, Y, and Z be the feet of perpendiculars from P to BC, CA, and AB, respectively.

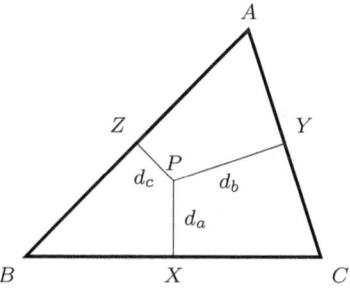

Because $PZ \perp AB$ and $PY \perp AC$, we find quadrilateral $AYPZ$ is cyclic. By the 2nd corollary of Ptolemy's theorem:

$$PA \cdot \sin A = YZ$$
$$= \sqrt{d_b^2 + d_c^2 - 2d_b d_c \cos(180° - \angle A)}$$
$$= \sqrt{d_b^2 + d_c^2 - 2d_b d_c \cos(\angle B + \angle C)}$$

Chapter 7: Solutions

Notice that:

$$d_b^2 + d_c^2 - 2d_b d_c \cos(\angle B + \angle C)$$
$$= d_b^2(\sin^2 \angle C + \cos^2 \angle C) + d_c^2(\sin^2 \angle B + \cos^2 \angle B)$$
$$\quad - 2d_b d_c(\cos \angle B \cos \angle C - \sin \angle B \sin \angle C)$$
$$= (d_b \sin \angle C + d_c \sin \angle B)^2 + (d_b \cos \angle C - d_c \cos \angle B)^2$$
$$\geq (d_b \sin \angle C + d_c \sin \angle B)^2$$

$$\therefore \quad PA \cdot \sin A \geq d_b \sin \angle C + d_c \sin \angle B \qquad (7.10)$$

Similarly,

$$PB \cdot \sin B \geq d_c \sin \angle A + d_a \sin \angle C \qquad (7.11)$$

$$PC \cdot \sin C \geq d_a \sin \angle B + d_b \sin \angle A \qquad (7.12)$$

These three above equations can derive

$$PA + PB + PC$$
$$= d_a \left(\frac{\sin \angle B}{\sin \angle C} + \frac{\sin \angle C}{\sin \angle B} \right) + d_b \left(\frac{\sin \angle C}{\sin \angle A} + \frac{\sin \angle A}{\sin \angle C} \right)$$
$$+ d_c \left(\frac{\sin \angle A}{\sin \angle B} + \frac{\sin \angle B}{\sin \angle A} \right)$$
$$\geq 2(d_a + d_b + d_c)$$

Equality holds if and only if $\sin \angle A = \sin \angle B = \sin \angle C$ and $d_a = d_b = d_c$.

Practice 8

Let P be a point inside $\triangle ABC$. Show that at least one of $\angle PAB$, $\angle PBC$, and $\angle PCA$ is less than or equal to $30°$.

(Ref 1991 IMO)

Chapter 7: Solutions

Suppose none of the three angles is less than or equal to 30°. Then none of them can be equal to or greater than 150°. Otherwise, if one of them is equal to or greater than 150°, then at least one of the two remaining angles must be equal to or less than 30°.

Let d_a, d_b, and d_c be the distances from P to BC, CA, and AB. Then we will have:

$$2d_a = 2PB\sin\angle PBC > (2\sin 30°)PB = PB$$

$$2d_b = 2PC\sin\angle PCA > (2\sin 30°)PC = PC$$

$$2d_c = 2PA\sin\angle PAB > (2\sin 30°)PA = PA$$

Adding these three relations above will result in contradiction to Erdös-Mordell inequality.

Practice 9

Let BD be the angle bisector of angle B in $\triangle ABC$ with D on side AC. The circumcircle of $\triangle BDC$ meets AB at E, while the circumcircle of $\triangle ABD$ meets BC at F. Prove that $AE = CF$.
(Ref 1996 St. Petersburg)

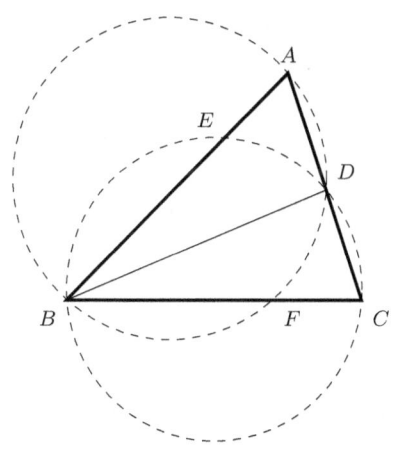

Chapter 7: Solutions

First, applying the power of a point theorem on circle $DEBC$ with respect to point A yields:

$$AE \cdot AB = AD \cdot AC$$

Next, applying the same theorem on circle $ADFB$ with respect to point C yields:

$$CF \cdot CB = CD \cdot CA$$

Therefore

$$\frac{AE}{CF} = \frac{AD}{DC} \cdot \frac{BC}{AB}$$

Because BD bisects $\angle B$, we find

$$\frac{AD}{DC} \cdot \frac{BC}{AB} = 1 \implies AE = CF$$

Practice 10

(**Miquel's Theorem**) In $\triangle ABC$, let points A', B', and C' locate on BC, CA, and AB, respectively. Show that the three circumcircles of $\triangle AB'C'$, $\triangle BC'A'$ and $\triangle CA'B'$ are concurrent. Its concurrent point is called the *Miquel point*.

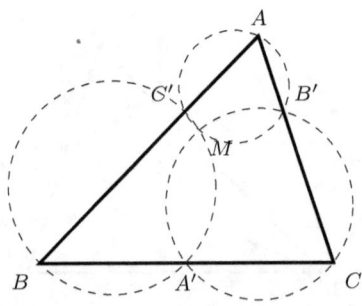

Let $\triangle AB'C'$'s and $\triangle BC'A'$'s circumcircles intersect at point M. The original claim is equivalent to showing that $M'A'CB'$ are concyclic.

Chapter 7: Solutions

Connect MA', MB' and MC'. Because $AB'MC'$ and $BC'MA'$ are concyclic, it must hold that

$$\angle B'MC' = 180° - \angle A \quad \text{and} \quad \angle C'MA' = 180° - \angle B$$

$\therefore \quad \angle A'MB' = 360° - (180° - \angle A) - (180° - \angle B) = 180° - \angle C$

Therefore, we find $CA'MB'$ are concyclic.

Practice 11

Let $ABCD$ be a circumscribed quadrilateral and O be its incenter. Show that

$$OA \cdot OC + OB \cdot OD = \sqrt{abcd}$$

where a, b, c, and d are lengths of its four sides.

First find the point E outside $ABCD$ such that $\triangle EAB \sim \triangle ODC$. This is obviously possible.

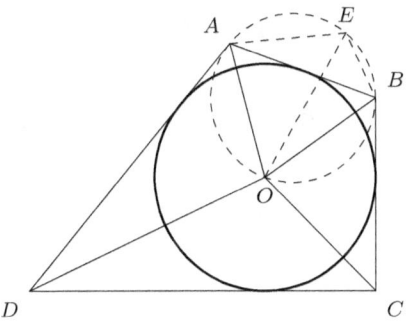

Next, by *Example 4.5.1 on page 34*, we find OA, OB, OC, and

Chapter 7: Solutions

OD bisect relevant interior angles. It follows that:

$$\angle AOB + \angle COD$$
$$=(180° - \frac{1}{2}(\angle A + \angle B)) + (180° - \frac{1}{2}(\angle C + \angle D))$$
$$=360° - \frac{1}{2}(\angle A + \angle B + \angle C + \angle D)$$
$$=180°$$

Because $\triangle EAB \sim \triangle ODC$, therefore $\angle E + \angle AOB = 180°$. This implies $EAOB$ are concyclic. By Ptolemy's theorem:

$$OE \cdot AB = OA \cdot BE + OB \cdot AE$$

Because $\frac{AE}{DO} = \frac{BE}{CO} = \frac{AB}{DC}$, the above relation is equivalent to

$$OE \cdot CD = OA \cdot OC + OB \cdot OD \qquad (7.13)$$

Next, because

$$\begin{cases} \angle AOE &= \angle ABE = \angle DCO = \angle OCB \\ \angle AEO &= \angle ABO = \angle OBC \end{cases}$$

We find $\triangle EAO \sim \triangle BOC$. Similarly, $\triangle EBO \sim \triangle AOD$. These mean that

$$\frac{OE}{BC} = \frac{OA}{OC} \quad \text{and} \quad \frac{OE}{AD} = \frac{BE}{OA} \implies \frac{OE^2}{BC \cdot AD} = \frac{BE}{OC}$$

But $\frac{BE}{OC} = \frac{AB}{CD}$, therefore

$$OE^2 \cdot CD = AB \cdot BC \cdot DA \implies OE^2 \cdot CD^2 = AB \cdot BC \cdot CD \cdot DA = abcd$$

Setting the above relation to *(7.13)* leads to the desired result.

7.5 Chapter 5

Practice 1

Show that a triangle's three inner angle bisectors are concurrent.

Let $\triangle ABC$'s three inner angle bisectors AD, BE, and CF meet their corresponding opposite sides at D, E, and F, respectively.

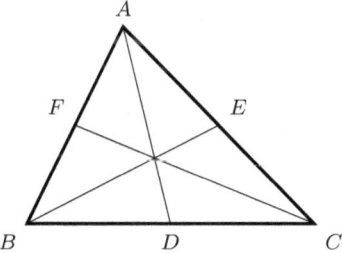

By the angle bisector theorem, we have

$$\frac{BD}{DC} = \frac{AB}{AC} \ , \quad \frac{CE}{EA} = \frac{BC}{BA} \quad \text{and} \quad \frac{AF}{FB} = \frac{CA}{CB}$$

$$\therefore \quad \frac{BD}{DC} \cdot \frac{CE}{EA} \cdot \frac{AF}{FB} = 1$$

By the converse of Ceva's theorem, we conclude that AD, BE, and CF are concurrent.

Practice 2

In $\triangle ABC$, let inner angle bisectors of $\angle B$ and $\angle C$ intersect their opposite sides at E and F, respectively. Let the exterior angle bisector of $\angle A$ intersect BC's extension at D. Show that $D, E,$ and F are collinear.[a]

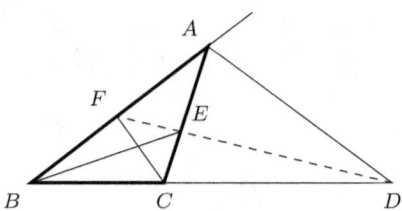

---

[a]This problem can also be solved by the physics method which is discussed in the book *Geometry Techniques*.

By the angle bisector theorems, we have

$$\frac{CE}{EA} = \frac{BC}{AB} \quad \frac{AF}{FB} = \frac{AC}{BC} \quad \frac{BD}{DC} = -\frac{AB}{AC}$$

Therefore

$$\frac{CE}{EA} \cdot \frac{AF}{FB} \cdot \frac{BD}{DC} = \frac{BC}{AB} \cdot \frac{AC}{BC} \cdot \left(-\frac{AB}{AC}\right) = -1$$

By the converse of Menelaus' theorem, we conclude $D, E,$ and F are collinear.

Practice 3

Given $\triangle ABC$, construct three squares outwards using its sides as bases, as shown. Point A_1, B_1, and C_1 are the midpoints of relevant sides. Show that AA_1, BB_1, and CC_1 are concurrent.

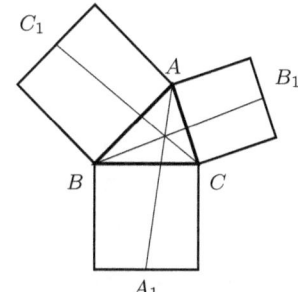

Let AA_1, BB_1, and CC_1 intersect BC, CA, and AB at A_2, B_2, and C_2, respectively. Connect BA_1, A_1C, CB_1, B_1A, AC_1 and C_1B.

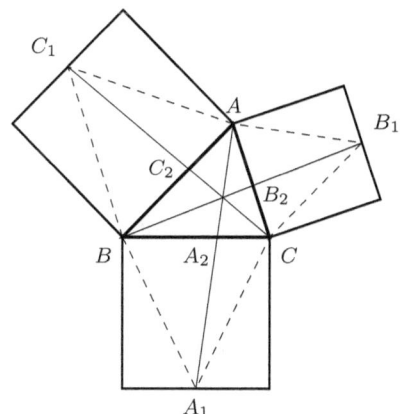

It is easy to see that
$$\angle CBA_1 = \angle BCA_1 = \angle ACB_1 = \angle CAB_1 = \angle ABC_1 = \angle BAC_1$$

Let them all equal to θ. Then[5]
$$\frac{BA_2}{CA_2} = \frac{S_{\triangle ABA_1}}{S_{\triangle ACA_1}} = \frac{AB \cdot BA_1 \cdot \sin(\angle B + \theta)/2}{AC \cdot CA_1 \cdot \sin(\angle C + \theta)/2} = \frac{AB \cdot \sin(\angle B + \theta)}{AC \cdot \sin(\angle C + \theta)}$$

---
[5] The side-angle-side formula for computing a triangle's area, $S = \frac{1}{2}ab\sin C$, will be discussed in *Section 6.1* on *page 49*.

Chapter 7: Solutions

Similarly, we shall have

$$\frac{CB_2}{B_2A} = \frac{BC \cdot \sin(\angle C + \theta)}{BA \cdot \sin(\angle A + \theta)}$$

and

$$\frac{AC_2}{C_2B} = \frac{CA \cdot \sin(\angle A + \theta)}{CB \cdot \sin(\angle B + \theta)}$$

Multiplying the above three equations leads to

$$\frac{CB_2}{B_2A} \cdot \frac{CB_2}{B_2A} \cdot \frac{AC_2}{C_2B} = 1$$

Therefore by the converse of Ceva's theorem, we find AA_1, BB_1, and CC_1 are concurrent.

Practice 4

Given $\triangle ABC$, construct three outward similar isosceles triangles using its sides as bases, $\triangle A_1BC$, $\triangle B_1CA$ and $\triangle C_1AB$, respectively. Show that AA_1, BB_1, and CC_1 are concurrent.

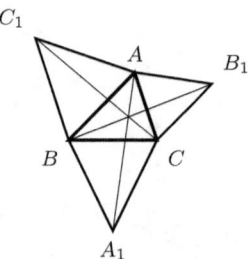

Without loss of generality, suppose $\angle A$ is $\triangle ABC$'s largest interior angle. Let the base angle of these isosceles triangles, e.g. $\angle C_1AB$, be θ. If $\angle A + \theta = 180°$, it is obvious that these three lines meet at A. We will prove the case when the sum is less than $180°$ here. The other case when the sum is greater than $180°$ can be proved in a similar way.

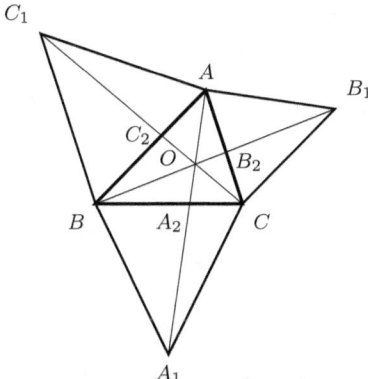

Let AA_1, BB_1, and CC_1 meet BC, CA, and AB at A_2, B_2, and C_2, respectively.

$$\frac{AC_2}{C_2B} = \frac{S_{\triangle ACC_1}}{S_{\triangle BCC_1}} = \frac{AC_1 \cdot CA\sin(\angle A + \theta)/2}{BC_1 \cdot BC\sin(\angle B + \theta)/2} = \frac{CA\sin(\angle A + \theta)}{BC\sin(\angle B + \theta)}$$

Similarly,
$$\frac{BA_2}{A_2C} = \frac{AB\sin(\angle B + \theta)}{CA\sin(\angle C + \theta)}$$

$$\frac{CB_2}{B_2A} = \frac{BC\sin(\angle C + \theta)}{AB\sin(\angle A + \theta)}$$

Multiplying these three equations results in

$$\frac{AC_2}{C_2B} \cdot \frac{BA_2}{A_2C} \cdot \frac{CB_2}{B_2A} = 1$$

By the converse of Ceve's theorem, AA_1, BB_1 and CC_1 are concurrent.

Chapter 7: Solutions

Practice 5

As shown in diagram below, find the degree measure of $\angle ADB$.

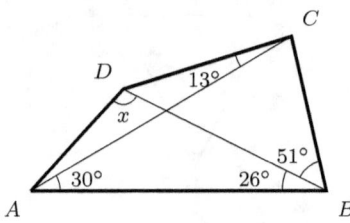

Let $\angle ADB = x$. It is easy to find $\angle ADC = x + 43°$.

Apply the Ceva's theorem (in the form of angles) on point A against $\triangle BDC$:

$$1 = \frac{\sin \angle DCA}{\sin \angle ACB} \cdot \frac{\sin \angle BDA}{\sin \angle ADC} \cdot \frac{\sin \angle CBA}{\sin \angle ABD} = \frac{\sin 13°}{\sin 73°} \cdot \frac{\sin x}{\sin (x + 43°)} \cdot \frac{\sin 77°}{\sin 26°}$$

It follows that:

$$\frac{\sin(x + 43°)}{\sin x} = \frac{\sin 13°}{\sin 73°} \cdot \frac{\sin 77°}{\sin 26°}$$
$$= \frac{\sin 13°}{\sin 73°} \cdot \frac{\cos 13°}{\sin 26°}$$
$$= \frac{1}{2 \sin 73°}$$
$$= \frac{\sin 30°}{\sin 73°}$$
$$= \frac{\sin 150°}{\sin 107°}$$
$$= \frac{\sin(107° + 43°)}{\sin 107°}$$

Therefore, we conclude $\angle ADB = x = 107°$.

Chapter 7: Solutions

Practice 6

The diagonals AC and CE of a regular hexagon $ABCDEF$ are divided by inner points M and N such that

$$AM : AC = CN : CE = r$$

Determine r if $B, M,$ and N are collinear.[a]

(Ref 1982 IMO)

---

[a]This problem can also be solved by using the coordinate method which is discussed in the book *Geometry Techniques*.

Join BE and let its intersection point with AC be P.

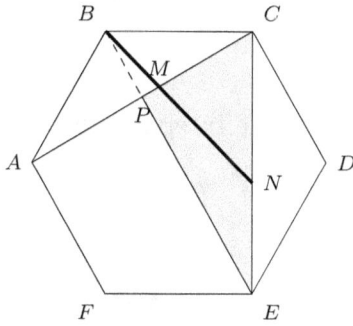

Apply Menelaus' theorem on $\triangle CPE$ with respect to line BMN:

$$\frac{CM}{MP} \cdot \frac{PB}{BE} \cdot \frac{EN}{NC} = -1 \qquad (7.14)$$

We note that:

(i) $CM : MP = 1 - r : r - \frac{1}{2}$

(ii) $PB : BE = -\frac{1}{4}$ (because $BP = AB \sin 30°$)

(iii) $BN : BC = 1 - r : r$

Setting $(i), (ii), (iii)$ to *Equation 7.14* leads to:

$$\frac{1-r}{r-\frac{1}{2}} \cdot \left(-\frac{1}{4}\right) \cdot \frac{1-r}{r} = -1 \implies r = \boxed{\frac{\sqrt{3}}{3}}$$

Practice 7

(Routh's Theorem) In $\triangle ABC$, let points D, E, and F be on BC, CA, and AB, respectively. If $AF : FB = x$, $BD : DC = y$, and $CE : EA = z$, then

$$S_{\triangle PQR} : S_{\triangle ABC} = \frac{(xyz - 1)^2}{(xy + y + 1)(yz + z + 1)(zx + x + 1)}$$

where P, Q, and R are the intersection points of AD, BE and CF, as shown.

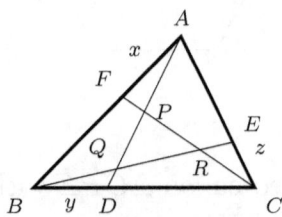

Without loss of generality, let's assume $S_{\triangle ABC} = 1$.

Applying Menelaus' theorem on $\triangle ABD$ with respect to line FC leads to (we only concern absolute values here.)

$$\frac{AF}{FB} \cdot \frac{BC}{CD} \cdot \frac{DP}{PA} = 1 \implies \frac{AP}{PD} = x(1+y) \implies \frac{AP}{AD} = \frac{x(1+y)}{1+x+xy}$$

It follows that

$$S_{\triangle APC} = \frac{AP}{AD} \cdot S_{\triangle ACD} = \frac{AP}{AD} \cdot \frac{DC}{BC} \cdot S_{\triangle ABC} = \frac{x}{1+x+xy}$$

Similarly, we can find

$$S_{\triangle BQA} = \frac{y}{1+y+yz} \quad \text{and} \quad S_{\triangle CRB} = \frac{z}{1+z+zx}$$

Therefore

$$S_{\triangle PQR} = 1 - \frac{x}{1+x+xy} - \frac{y}{1+y+yz} - \frac{z}{1+z+zx}$$
$$= \frac{(xyz-1)^2}{(1+x+xy)(1+y+yz)(1+z+zx)}$$

Practice 8

(Pascal's Theorem) Let $A, B, C, D, E,$ and F be points on a circle (not necessarily in cyclic order). Let AB and DE meet at P, BC and EF meet at Q, CD ad FA meet at R. Prove P, Q, and R are collinear.

Let BC and AF meet at X, DE and AF meet at Y, BC and DE meet at Z. The objective is to show that in $\triangle XYZ$, the following relation holds:

$$\frac{XR}{RY} \cdot \frac{YP}{PZ} \cdot \frac{ZQ}{QX} = -1 \qquad (7.15)$$

If so, then by the converse of Menelaus' theorem, we can assert $P, Q,$ and R are collinear.

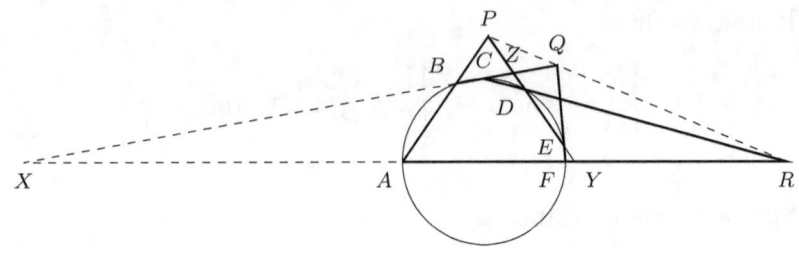

In order to show *(7.15)*, apply Menelaus' theorem on $\triangle XYZ$ by the following transversals:

$$PA \implies \frac{XA}{AY} \cdot \frac{YP}{PZ} \cdot \frac{ZB}{BX} = -1$$
$$QF \implies \frac{XF}{FY} \cdot \frac{YE}{EZ} \cdot \frac{ZQ}{QX} = -1$$
$$RC \implies \frac{XR}{RY} \cdot \frac{YD}{DZ} \cdot \frac{ZC}{CX} = -1$$

Then by the Power of Point theorem, we have

$$XA \cdot XF = XB \cdot XC \implies \frac{XA \cdot XF}{XB \cdot XC} = 1$$
$$YE \cdot YD = YF \cdot YA \implies \frac{YD \cdot YE}{YA \cdot YF} = 1$$
$$ZC \cdot ZB = ZD \cdot ZE \implies \frac{ZB \cdot ZC}{ZD \cdot ZE} = 1$$

Multiplying the above six equations yields *(7.15)* immediately.

Chapter 7: Solutions

Practice 9

(**Newton's Theorem**) A circle is inscribed in quadrilateral $ABCD$ with sides touch the circle at E, F, G, and H, as shown. Prove AC, BD, EG, and FH are concurrent.

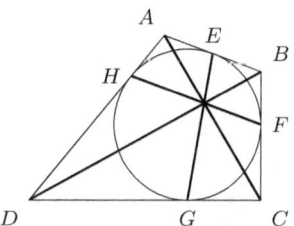

Let EH and FG meet at point X, EG and FH meet at point O.

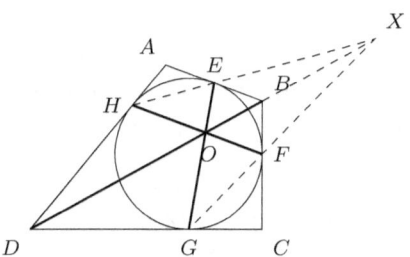

By applying Pascal's theorem (see previous practice) on points H, F, F, G, E, E, we find O, B, and X are collinear. (Because both E and F are tangent points, lines EE and FF will be the two tangent lines which intersect at point B.)

Similarly, applying Pascal's theorem on E, G, G, F, H, H, we can show D, O, and B are collinear. Therefore, B, O, and D must be collinear. This means that lines EG, FH, and BD are concurrent at O. By the same reasoning EG, FH, and AC are concurrent at O. Then Newton's theorem follows.

Practice 10

(Desargues' Theorem) Given two triangles ABC and $A'B'C'$. Suppose that lines AA', BB', and CC' are concurrent. Let AB and $A'B'$, BC and $B'C'$, CA and $C'A$ intersect at X, Y, and Z, respectively. Show that X, Y, Z are collinear.

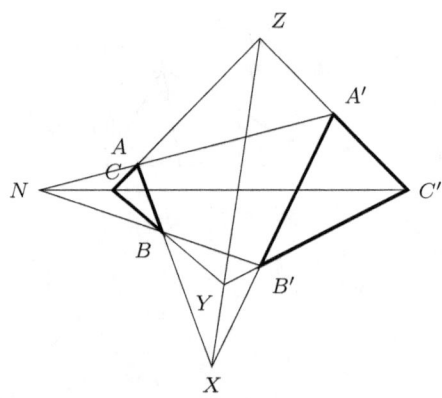

Because X, Y, and Z are points on the extensions of $\triangle ABC$'s three sides, it is sufficient to show that

$$\frac{AX}{XB} \cdot \frac{BY}{YC} \cdot \frac{CZ}{ZA} = -1$$

First, consider $\triangle NAB$ and its transversal $A'B'X$. By Menelaus' theorem:

$$\frac{AX}{XB} \cdot \frac{BB'}{B'N} \cdot \frac{NA'}{A'A} = -1$$

Next, consider $\triangle NAC$ and its transversal $ZA'C'$:

$$\frac{CZ}{ZA} \cdot \frac{AA'}{A'N} \cdot \frac{NC'}{C'C} = -1$$

Finally, consider $\triangle NBC$ and its transversal $YB'C'$:

$$\frac{NB'}{B'B} \cdot \frac{BY}{YC} \cdot \frac{CC'}{C'N} = -1$$

Now, multiplying these three equations yields the desired result.

Chapter 7: Solutions

7.6 Chapter 6

Practice 1

Give an outline of proof for the Brahmagupta's formula using Heron's formula and similar triangles.

If the two pairs of opposite sides in quadrilateral $ABCD$ are both parallel to each other, $ABCD$ will be a rectangle. This means $a = c$, $b = d$, and $p = a + b = c + d$. Consequently,

$$S = \sqrt{(p-a)(p-b)(p-c)(p-d)} = \sqrt{b \cdot a \cdot d \cdot c} = ab$$

This obviously holds.

Otherwise, without loss of generality, let's assume AD and BC intersects at point P as shown below.

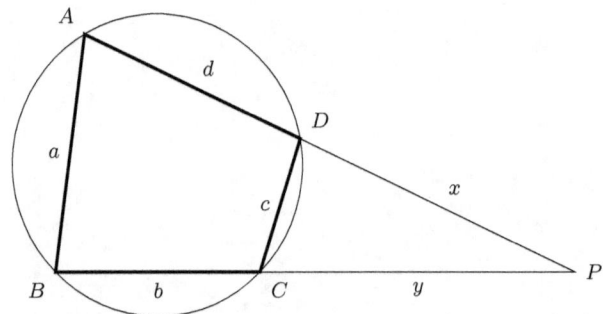

Because $ABCD$ is cyclic, it must hold that $\triangle PCD \sim \triangle PAB$ by the AAA rule. Therefore we have:

$$\frac{x}{b+y} = \frac{y}{x+d} = \frac{c}{a}$$

Solving this system leads to

$$x = \frac{abc + c^2 d}{a^2 - c^2} \quad \text{and} \quad y = \frac{acd + bc^2}{a^2 - c^2} \tag{7.16}$$

or

$$b + y = \frac{a(ab + cd)}{a^2 - c^2} \quad \text{and} \quad d + x = \frac{a(bc + ad)}{a^2 - c^2}$$

Meanwhile, $S_{\triangle PCD} : S_{\triangle PBA} = (\frac{c}{a})^2$. Therefore

$$S = S_{ABCD} = \frac{a^2 - c^2}{a^2} \cdot S_{\triangle PAB}$$

Notice that the three sides of $\triangle PAB$ are all known: a, $\frac{a(ab+cd)}{a^2-c^2}$, and $\frac{a(bc+ad)}{a^2-c^2}$. It is then possible to compute its area using Heron's formula from which we can derive S.

Practice 2

Derive Brahmagupta's formula (area of a cyclic quadrilateral, *page 55*) from Bretschneider's formula (area of any quadrilateral, *page 55*).

Let a, b, c, d be the edges' lengths in that order and m, n be the diagonals' lengths, Bretschneider's formula states

$$S = \frac{1}{4}\sqrt{4m^2 n^2 - (a^2 + c^2 - b^2 - d^2)}$$

When the quadrilateral is cyclic, Ptolemy's theorem states

$$mm = ac + bd$$

Chapter 7: Solutions

Therefore
$$\begin{aligned} S &= \frac{1}{4}\sqrt{4m^2n^2 - (a^2 + c^2 - b^2 - d^2)} \\ &= \frac{1}{4}\sqrt{4(ac+bd)^2 - (a^2 + c^2 - b^2 - d^2)} \\ &= \frac{1}{4}\sqrt{-a^4 - b^4 - c^4 - d^4 + 2a^2b^2 + 2a^2c^2 + \cdots} \\ &= \frac{1}{4}\sqrt{(-a+b+c+d)(a-b+c+d)(a+b-c+d)(a+b+c-d)} \end{aligned}$$

Let $p = \frac{a+b+c+d}{2}$. The above formula can be rewritten as
$$S = \sqrt{(p-a)(p-b)(p-c)(p-d)}$$
which is Brahmagupta's formula.

Practice 3

(**Subtended Angle Theorem**) Let $\angle BAD = \alpha$ and $\angle CAD = \beta$. Show that point D is on BC if and only if the following relation holds:

$$\frac{\sin(\alpha + \beta)}{AD} = \frac{\sin \alpha}{AC} + \frac{\sin \beta}{AB}$$

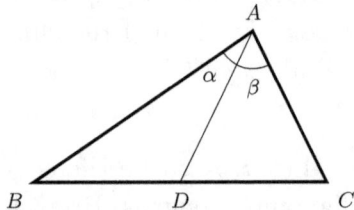

Point D locates on BC is equivalent to
$$S_{\triangle ABC} = S_{\triangle ABD} + S_{\triangle ACD}$$
$$\frac{1}{2} \cdot AB \cdot AC \cdot \sin(\alpha + \beta) = \frac{1}{2} \cdot AB \cdot AD \cdot \sin \alpha + \frac{1}{2} \cdot AC \cdot AD \cdot \sin \beta$$

$$\frac{\sin(\alpha + \beta)}{AD} = \frac{\sin \alpha}{AC} + \frac{\sin \beta}{AB}$$

Chapter 7: Solutions

Practice 4

(Steiner's Theorem) Let M be a point inside $\triangle ABC$ satisfying
$$\frac{\sin \angle BMC}{a} = \frac{\sin \angle CMA}{b} = \frac{\sin \angle CMA}{c}$$
Show that, for any point P inside $\triangle ABC$, the following relation always holds:
$$a \cdot PA + b \cdot PB + c \cdot PC \geq a \cdot MA + b \cdot MB + c \cdot MC$$

The equality holds if and only if M and P coincide.

(Note: In fact, the condition that P is inside $\triangle ABC$ is unnecessary.)

Draw $A'B' \perp MC$, $B'C' \perp MA$, and $C'A' \perp MB$, as shown. It follows that
$$\angle BMC + \angle A' = \angle CMA + \angle B' = \angle AMB = \angle C' = 180°$$

or

$$\sin \angle BMC = \sin \angle A', \ \sin \angle CMA = \sin \angle B', \ \sin \angle AMB = \sin \angle C'$$

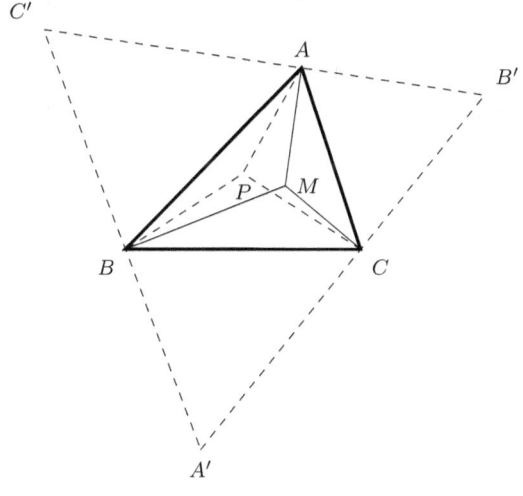

Chapter 7: Solutions

Applying the Law of Sines and the given conditions leads to

$$\frac{A'B'}{AB} = \frac{B'C'}{BC} = \frac{C'A'}{CA} = k$$

where k is a positive constant.

$$\begin{aligned}
S_{\triangle A'B'C'} &= S_{\triangle MA'B'} + S_{\triangle MB'C'} + S_{\triangle MC'A'} \\
&= \frac{1}{2} \times (A'B' \cdot MC + B'C' \cdot MA + C'A' \cdot MB) \\
&= \frac{k}{2} \times (c \cdot MC + a \cdot MA + b \cdot MB)
\end{aligned}$$

Meanwhile,

$$\begin{aligned}
&\frac{k}{2} \times (c \cdot PC + a \cdot PA + b \cdot PB) \\
&= \frac{1}{2} \times (A'B' \cdot PC + B'C' \cdot PA + C'A' \cdot PB)) \\
&\geq S_{\triangle MA'B'} + S_{\triangle MB'C'} + S_{\triangle MC'A'} \\
&= S_{\triangle A'B'C'}
\end{aligned}$$

Therefore

$$\frac{k}{2} \times (c \cdot PC + a \cdot PA + b \cdot PB) \geq \frac{k}{2} \times (c \cdot MC + a \cdot MA + b \cdot MB)$$

Equality holds if and only if $PA \perp B'C'$, $PB \perp C'A'$, and $PC \perp A'B'$, or equivalently, P and M coincide.

Chapter 7: Solutions

Practice 5

(**Finsler-Hadwiger's Inequalities**) Prove: in any triangle ABC, the following inequalities hold:

$$4\sqrt{3} \cdot S + Q \leq a^2 + b^2 + c^2 \leq 4\sqrt{3} \cdot S + 3Q$$

where S is the triangle's area and $Q = (a-b)^2 + (b-c)^2 + (c-a)^2$.

Tip: By the Schur inequality, we can show that for any positive real number x, y, and z, the following inequality always holds. You may use this to solve this problem.

$$9xzy \geq (x+y+z)(2xy + 2yz + 2zx - x^2 - y^2 - z^2) \qquad (7.17)$$

Let a, b, and c be the lengths of triangle's three sides. Then there exist three positive real numbers x, y, and z such that $a = y+z$, $b = z+x$, and $c = x+y$. By the Heron's formula, we have

$$S = \sqrt{xyz(x+y+z)}$$

The to-be-proved relations are equivalent to

$$a^2 + b^2 + c^2 - 3Q \leq 4\sqrt{3} \cdot S \leq a^2 + b^2 + c^2 - Q$$

The left side is equivalent to

$$(y+z)^2 + (z+x)^2 + (x+y)^2 - 3(x-y)^2 - 3(y-z)^2 - 3(z-x)^2$$
$$\leq 4\sqrt{xyz(x+y+z)}$$
$$\Leftrightarrow 2xy + 2yz + 2zx - x^2 - y^2 - z^2 \leq \sqrt{3xyz(x+y+z)} \qquad (7.18)$$

From *(7.17)*, we have

$$2xy + 2yz + 2zx - x^2 - y^2 - z^2 \leq \frac{9xyz}{x+y+z}$$

$$\Leftrightarrow 27xyz \leq (x+y+z)^3$$

$$\Leftrightarrow \sqrt[3]{xyz} \le \frac{x+y+z}{3}$$

The last inequality obviously holds by the AGM inequality. Therefore *(7.18)* holds.

Next, let's prove the right side:
$$4\sqrt{3} \cdot S \le a^2 + b^2 + c^2 - Q$$

This is equivalent to
$$4\sqrt{3xyz(x+y+z)} \le$$
$$(y+z)^2 + (z+x)^2 + (x+y)^2 - (x-y)^2 - (y-z)^2 - (z-x)^2$$

$$\Leftrightarrow \sqrt{3xyz(x+y+z)} \le xy + yz + zx$$
$$\Leftrightarrow 3xyz(x+y+z) \le (xy+yz+zx)^2$$
$$\Leftrightarrow (xy)(xz) + (xy)(yz) + (xz)(yz) \le (xy)^2 + (xz)^2 + (zx)^2$$

The last relation holds because, for any positive real numbers p, q, and r, we have
$$(p-q)^2 + (q-r)^2 + (r-p)^2 \ge 0$$
$$p^2 - 2pq + q^2 + q^2 - 2qr + r^2 + r^2 - 2rp + p^2 \ge 0$$
$$p^2 + q^2 + r^2 \ge pq + qr + rp$$

Practice 6

(Wetizenbock's Inequality) In any triangle ABC, the following relation holds:
$$a^2 + b^2 + c^2 \ge 4\sqrt{3} \cdot S$$

This is obvious by the *Finsler-Hadwiger's Inequalities*. A direct proof is given in the book *Geometry Techniques*.

Chapter 7: Solutions

Practice 7

Does there exist a point P inside $\triangle ABC$ such that any line passing P will divide $\triangle ABC$ into two equal areas?
(Ref 1998 China)

The answer is No.

Suppose there exists such a point P. Let's connect AP and also extend it to meet BC at D. Because $S_{\triangle ABD} = S_{\triangle ACD}$, it must hold that D is the midpoint of BC. This implies P lies on the median from A. Similarly, P must locate on other medians too. This leads to the conclusion that P is the centroid.

Now let's draw a line which is parallel to BC and passes point P. Suppose this line intersects AB and AC at M and N, respectively. It is easy to compute

$$S_{\triangle AMN} = \left(\frac{2}{3}\right)^2 S_{\triangle ABC} \neq \frac{1}{2} S_{\triangle ABC}$$

Practice 8

Let G be the centroid of $\triangle ABC$. A line passing through G intersects AB and AC at P and Q, respectively. Show that

$$\frac{AB}{AP} + \frac{AC}{AQ} = 3$$

This relation can be proved by setting $BM = CM$ and $AM : AN = 3 : 2$ in *(6.10)* on *page 54*.

Chapter 7: Solutions

Practice 9

Let H be the orthocenter of a non-right triangle ABC. A line passing H intersects AB and AC at P and Q, respectively. Show that

$$\frac{AB}{AP} \cdot \tan \angle B + \frac{AC}{AQ} \cdot \tan \angle C = \tan \angle A + \tan \angle B + \tan \angle C$$

Let's first observe the diagram below (this diagram is not the same as the one associated with the original problem):

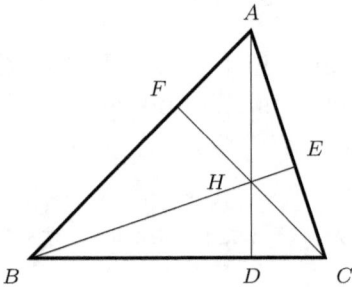

$$Rt\triangle AHE \sim Rt\triangle BCE \implies \tan \angle A = \frac{BE}{AE} = \frac{BC}{AH}$$

$$\therefore \quad \frac{AD}{AH} = \frac{AD \cdot \tan \angle A}{BC}$$

Meanwhile,

$$\because \quad \tan \angle B = \frac{AD}{BD} \quad \text{and} \quad \tan \angle C = \frac{AD}{DC}$$

$$\therefore \quad \frac{BD}{BC} = \frac{AD}{BC \cdot \tan \angle B} \quad \text{and} \quad \frac{CD}{BC} = \frac{AD}{BC \cdot \tan \angle C}$$

Chapter 7: Solutions

Setting these relations to *(6.10)* on *page 54*:

$$\frac{AD}{AH} = \frac{AB}{AP} \cdot \frac{CD}{BC} + \frac{AC}{AQ} \cdot \frac{BD}{BC}$$

$$\frac{AD \cdot \tan \angle A}{BC} = \frac{AB}{AP} \cdot \frac{AD}{BC \cdot \tan \angle C} + \frac{AC}{AQ} \cdot \frac{AD}{BC \cdot \tan \angle B}$$

$$\tan \angle A = \frac{AC}{AQ} \cdot \frac{1}{\tan \angle B} + \frac{AB}{AQ} \cdot \frac{1}{\tan \angle C}$$

Multiplying $(\tan \angle B \cdot \tan \angle C)$ on both sides and then applying the following trigonometric identity will yield the to-be-proved claim immediately.

$$\frac{AB}{AP} \cdot \tan \angle B + \frac{AC}{AQ} \cdot \tan \angle C = \tan \angle A + \tan \angle B + \tan \angle C$$

Practice 10

(Japanese Theorem for Cyclic Quadrilaterals) Let $ABCD$ be a cyclic quadrilateral. Prove that the incenters of $\triangle ABC$, $\triangle BCD$, $\triangle CDA$, and $\triangle DAB$ form a rectangle.

Let J, K, L, and M be the four incenters. Without loss of generality, let's show that $\angle LMJ$ is right.

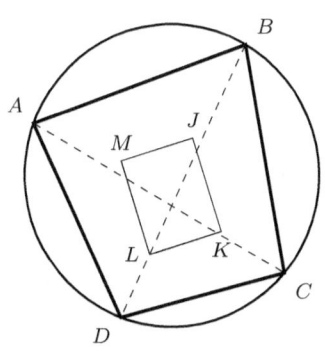

127

In $\triangle ABD$, applying *Example 6.2.5* on *page 63* leads to

$$\angle AMD = 90° + \frac{1}{2}\angle ABD$$

Similarly, in $\triangle ACD$, we have

$$\angle ALD = 90° + \frac{1}{2}\angle ACD$$

Because $ABCD$ is cyclic, it must hold that $\angle ABD = \angle ACD$, or equivalently, $\angle AMD = \angle ALD$. This implies $AMLD$ is cyclic. It follows that

$$\angle AML = 180° - \angle ADL = 180° - \frac{1}{2}\angle ADC$$

By a similar reasoning, we can derive

$$\angle AMJ = 180° - \frac{1}{2}\angle ABC$$

Adding the last two relations yields

$$\angle AML + \angle AMJ = 360° - \frac{1}{2} \times (\angle ADC + \angle ABC) = 270°$$

This concludes that $\angle LMJ = 90°$.

Practice 11

Let H be the orthocenter of $\triangle ABC$. Extend AH, BH, CH, and let them meet opposite sides at D, E, and F, respectively. Show that

$$AH \cdot HD = BH \cdot HE = CH \cdot HF$$

Chapter 7: Solutions

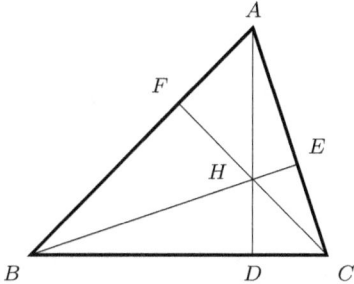

$\triangle AHE \sim \triangle BHD \implies \dfrac{AH}{HE} = \dfrac{BH}{HD} \implies AH \cdot HD = BH \cdot HE$

$\triangle BHF \sim \triangle CHE \implies \dfrac{BH}{HF} = \dfrac{CH}{HE} \implies BH \cdot HE = CH \cdot HF$

$\therefore \quad AH \cdot HD = BH \cdot HE = CH \cdot HF$

Practice 12

Let H be $\triangle ABC$'s orthocenter and R be its circumradius. Show that
$$\dfrac{AH}{|\cos \angle A|} = \dfrac{BH}{|\cos \angle B|} = \dfrac{CH}{|\cos \angle C|} = 2R$$

Let's prove the conclusion when $\triangle ABC$ is acute here. Other cases can be proved in a similar way.

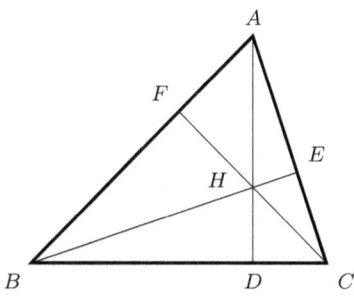

Chapter 7: Solutions

First,

$$\angle C = \angle AHE \implies \sin \angle C = \sin \angle AHE = \frac{AE}{AH}$$

Next, in $Rt\triangle ABE$:

$$AE = AB \cdot \cos \angle A \implies AH = \frac{AB \cdot \cos \angle A}{\sin \angle C} = 2R \cdot \cos \angle A$$

Finally, because $\angle A$ is acute, the last relation implies

$$\frac{AH}{|\cos A|} = 2R$$

The other two relations can be derived in a similar way.

Practice 13

Given a triangle, show that the distance between an vertex and its orthocenter is twice of the distance between its circumcenter to its opposite side.

Let H and O be the orthocenter and circumcenter of $\triangle ABC$ as shown.

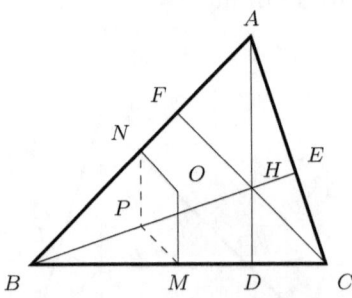

Let M, N, and P be the midpoints of BC, BA, and BH, respectively. Then, it must hold that

$$NP \parallel AH \parallel OM \quad \text{and} \quad MP \parallel CH \parallel ON$$

because $OM \perp BC$, $AH \perp BC$, $ON \perp AB$, and $CH \perp AB$.

Therefore $NPOM$ is a parallelogram. Furthermore, because P and N are midpoints of BH and AB, respectively, it must be true that $AH = 2PN$, equivalently, $AH = 2OM$ because $NP = OM$.

Practice 14

(**Euler's Line**) Let O, G, and H, be the circumcenter, the centroid, and the orthocenter of $\triangle ABC$, respectively. Show that they are collinear and, in addition, $GH = 2OG$.

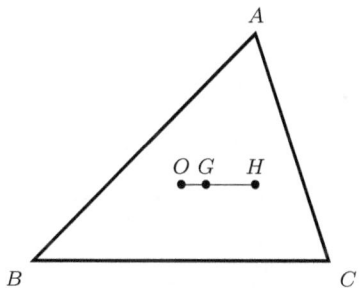

Let D be the midpoint of BC. Connect DO, AH, AG, DG, DO, and GH.

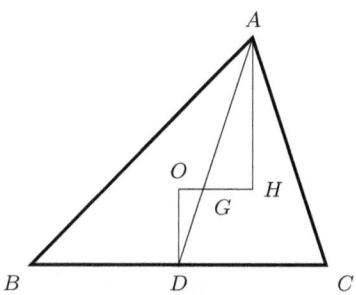

131

Chapter 7: Solutions

Because G is the centroid and D is the midpoint of BC, G must locate on AD and $AG : GD = 2 : 1$. Meanwhile, the triangle formed by midpoints of $\triangle ABC$ is similar to the original triangle ABC with a scaling factor of 0.5 (see *Section 6.2.3* on *page 61*). Consequently, we know $AH : DO = 2 : 1$ because they are corresponding segments in two similar triangles. It is obvious that $\angle ODG = \angle HAG$ because $AH \parallel OD$. Consequently we find $\triangle AHG \sim \triangle DOG$. This means $\angle AGH = \angle DGO$, or OGH are collinear, and $HG : GO = 2 : 1$.

Practice 15

Two mutually tangent congruent circles are internally tangent to a $5 - 12 - 13$ triangle, as shown. Find the radius of these two circles.

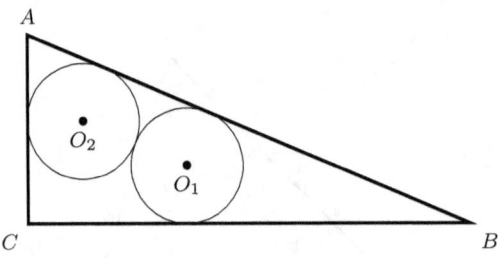

Let O be the incenter of $\triangle ABC$. Connect AO, BO and O_1O_2.

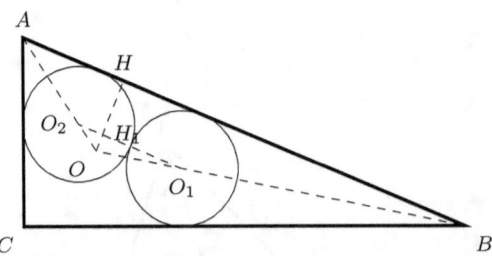

Because circle O_1 is tangent to AB and BC, its center O_1 must lie on the angle bisector of angle B, or BO. Similarly, O_2 must be on AO.

Because both circles are tangle to side AB, their centers must be equidistant to AB. This implies $O_1O_2 \parallel AB$ which leads to $\triangle OO_1O2 \sim \triangle OBA$.

Let the desired radius be r. Draw $OH \perp AB$ meeting AB at H and O_1O_2 at H_1. Clearly, OH and OH_1 are altitudes of $\triangle OO_1O_2$ and $\triangle OBA$, respectively. Therefore

$$\triangle OO_1O2 \sim \triangle OBA \implies \frac{O_1O_2}{BA} = \frac{OH_1}{OH}$$

It is easy to see that OH is $\triangle ABC$'s inradius which equals 2 by applying *Formula 2.3* on *page 7*. HH_1 must equal r. Hence $OH_1 = OH - H_1H = 2 - r$.

$$\therefore \frac{O_1O_2}{BA} = \frac{OH_1}{OH} \Leftrightarrow \frac{2r}{13} = \frac{2-r}{2} \implies r = \boxed{\frac{26}{17}}$$

Practice 16

Let G and I be the centroid and incenter of $\triangle ABC$, respectively. If $GI \parallel BC$, show that $AB + AC = 2 \cdot BC$.

Connect and extend AG and AI such that they intersect BC at M and T, respectively.

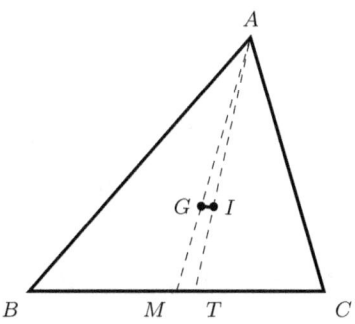

Because G is the centroid and $GI \parallel BC$, it must hold that
$$AI : TI = AG : MG = 2$$

Now connect CI. Because I is the incenter, CI must bisect $\angle C$. By the angle bisector theorem:
$$\frac{AC}{CT} = \frac{AI}{TI} = 2 \implies AC = 2 \cdot CT$$

Similarly, it must hold that $AB = 2 \cdot BT$. Therefore
$$AB + AC = 2 \cdot (BT + CT) = 2 \cdot BC$$

Practice 17

Let O be the circumcenter of an acute $\triangle ABC$. If point H lies inside $\triangle ABC$ and satisfies $\angle BAO = \angle HAC$, $\angle ABO = \angle HBC$, show that H is the orthocenter.

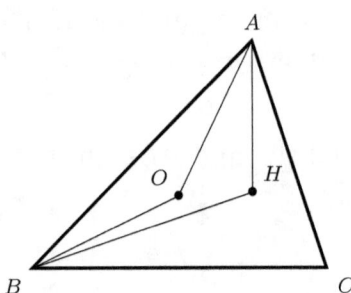

$$\angle HAC = \angle BAO = 90° - \frac{1}{2}\angle AOB = 90° - \angle C$$
$$\therefore \quad AH \perp BC$$

Similarly, we can show that $BH \perp AC$. Therefore H must be the orthocenter.

Index

Angle Bisector Theorem, 13
Apollonius' Theorem, 20, 83

Brahmagupta Theorem, 35, 93
Brahmagupta's Formula, 55, 65, 118, 119
Bretschneider's Formula, 55
British Flag Theorem, 10, 75
Butterfly Theorem, 36, 96

Carnot's Theorem, 36, 94
Central Angle Theorem, 23
Ceva's Theorem, 40
Ceva's Theorem - Trigonometric Form, 42

De Gua's Theorem, 4, 11, 76
Desargues' Theorem, 48, 116

Erdös-Mordell Inequality, 37, 99
Euler's Line, 64, 69, 131

Fermat Point, 64
Finsler-Hadwiger's Inequalities, 66, 123

Gemoetry Mean Theorem, 7

Heron's Formula, 50

Intersecting Chords Theorem, 31
Intersecting Secants Theorem, 31

INDEX

Japanese Theorem for Cyclic Quadrilaterals, 67, 127

Law of Cosines, 16
Law of Cosines (3-Dimensional Space), 22, 89
Law of Sines, 15
Law of Tangents, 21, 86

Menelaus' Theorem, 42
Miquel's Theorem, 38, 102

Napoleon's Triangle, 29
Newton's Theorem, 47, 115
Nine-point Circle, 64

Pascal's Theorem, 47, 113
Pick's Theorem, 56
Power of a Point Theorem, 31
    Intersecting Chords Theorem, 31
    Intersecting Secants Theorem, 31
Ptolemy's Theorem, 30
    Ptolemy's Inequality, 30
Pythagorean Theorem, 3
Pythagorean Triplet Formula, 5

Ratio of Triangle Areas, 52
Routh's Theorem, 46, 112

Steiner's Theorem, 66, 121
Stewart Theorem, 18
Subtended Angle Theorem, 65, 120

Wetizenbock's Inequaility, 66, 124

www.ingramcontent.com/pod-product-compliance
Lightning Source LLC
Chambersburg PA
CBHW071442180526
45170CB00001B/424